SpringerBriefs in Computer Science

SpringerBriefs present concise summaries of cutting-edge research and practical applications across a wide spectrum of fields. Featuring compact volumes of 50 to 125 pages, the series covers a range of content from professional to academic.

Typical topics might include:

- A timely report of state-of-the art analytical techniques
- A bridge between new research results, as published in journal articles, and a contextual literature review
- A snapshot of a hot or emerging topic
- An in-depth case study or clinical example
- A presentation of core concepts that students must understand in order to make independent contributions.

Briefs allow authors to present their ideas and readers to absorb them with minimal time investment. Briefs will be published as part of Springer's eBook collection, with millions of users worldwide. In addition, Briefs will be available for individual print and electronic purchase. Briefs are characterized by fast, global electronic dissemination, standard publishing contracts, easy-to-use manuscript preparation and formatting guidelines, and expedited production schedules. We aim for publication 8–12 weeks after acceptance. Both solicited and unsolicited manuscripts are considered for publication in this series.

**Indexing: This series is indexed in Scopus, Ei-Compendex, and zbMATH **

Herkulaas Combrink

Computational Infodemiology

Belief System Mapping and Social Stress

Springer

Herkulaas Combrink
Economics and Management Sciences
and Interdisciplinary Centre for Digital
Futures
University of the Free State
Bloemfontein, South Africa

ISSN 2191-5768 ISSN 2191-5776 (electronic)
SpringerBriefs in Computer Science
ISBN 978-3-032-10129-7 ISBN 978-3-032-10130-3 (eBook)
https://doi.org/10.1007/978-3-032-10130-3

This Springer imprint is published by the registered company Springer Nature Switzerland AG
The registered company address is: Gewerbestrasse 11, 6330 Cham, Switzerland

It is not about what education can do for me, but what I can do for others through my education.

—Dr. Herkulaas Combrink

Preface I

We live in a paradoxical time, never has so much information been so readily available, and yet never has collective understanding felt so fragile. From social media algorithms to anonymous misinformation campaigns, the very architecture of our digital world is shaping what, and who we believe. In this book, we are offered not only a diagnosis of the systemic dynamics that drive infodemics but also a set of theoretical, computational, and belief maps that drive these infodemics.

As someone who has worked in computer science, data science, health, and complex science, Computational Infodemiology emerges as a field that encapsulates all these disciplines. The work presented here does something rare, it formalises belief systems into computationally traceable structures, while simultaneously never losing sight of their social, emotional, and political realities.

We are invited to view infodemic risk not as individual vulnerability, but as an emergent property of systems, systems which can be understood, simulated, and ultimately transformed. What distinguishes this work is its insistence on nuance. The simulations presented show just how sensitive these systems are to variables like uncertainty, misinformation flow, and fact-checking efficacy. In doing so, the book makes a powerful case:

> If belief systems can be modelled, they can be governed, not through coercion, but also through structural design.

For researchers, technologists, policymakers, and public health leaders, this book will serve as both roadmap and provocation. It challenges us to move beyond reactive misinformation strategies and instead rethink the very structure of our information environments. It demands that we take seriously the idea that information is not neutral, and that the design of digital systems must be approached with the same ethical rigour we apply to public infrastructure.

The emergence of Computational Infodemiology marks a turning point in how we understand the role of data in public discourse. And this book is, without doubt, a cornerstone in that evolution. It gives shape to a field that is urgently needed, and it does so with intellectual rigour, systems thinking, and clarity.

May it be read widely, debated fiercely, and used constructively. The future of informed society may very well depend on how deeply we understand the lessons within these pages.

Bloemfontein, South Africa
July 2025

Dr. Herkulaas Combrink

Preface II

This book was born out of a growing sense that the dynamics of misinformation are no longer incidental to public life. Across health crises, elections, social movements, and scientific discourse, the influence of belief systems, formed, fractured, and amplified in digital environments has become central to the way society's function or falter. As I engaged with these challenges, particularly from the vantage point of countries where the digital divide is closing, I became convinced that we need not only better data or stronger fact-checking but a new epistemology for the information age. This is the core motivation behind Computational Infodemiology. Over the past few years, I have worked to shape this emerging field as one that combines computational modelling, reinforcement learning, and natural language processing with deep sociological insight. This book represents the synthesis of that effort. It explores not only how infodemics spread, but also how belief systems evolve under pressure, and how we might simulate, and map in those processes using agent-based and data-driven methods.

Chapter 1 introduces the reader to the conceptual terrain of infodemics, tracing their sociotechnical roots in algorithmic feedback loops and uncertainty amplification. Drawing on frameworks such as systemic intervention, social norm formation, and Nussbaum's spheres of influence, it argues that belief systems are not simply individual cognitive processes, but are embedded in social, technological, and ethical infrastructures.

Chapter 2 provides the computational foundation, introducing agent-based modelling (ABM) and reinforcement learning (RL) as methodological tools to capture the adaptive, decentralised, and often chaotic nature of information spread. Particular attention is given to network topologies, which shape how information circulates and stabilises.

Chapter 3 introduces the novel concept of a belief system map, offering a structured framework for identifying, categorising, and visualising belief states across agents. This approach allows us to move beyond binary notions of "true" and "false" to examine how beliefs gain traction, how they conflict, and how agents transition between states.

In Chap. 4, the theoretical and computational tools come together through a series of simulations. These span baseline models, conflict-driven dynamics, and community-specific behaviours, revealing the conditions under which beliefs stabilise, polarise, or fragment. From deterministic convergence to chaotic oscillations and sceptic drop-off risks, these models allow us to study belief ecologies with clarity and nuance.

Chapter 5 turns to the development of the Social Stress Indicator (SSI), a composite metric designed to detect and quantify latent social tension during periods of heightened informational conflict. Drawing on Search volume indices, sentiment analysis, and topic modelling, this chapter builds a multi-layered approach to infoveillance, offering a practical tool for early detection of belief shocks, digital unrest, and infodemic surges. This work is as much a systems thinking exercise as it is a technical one. It asks: *What happens when misinformation is rewarded and scepticism is structurally penalised? How can we design information environments where truth is not only accessible, but strategically advantageous?* And perhaps most urgently, *What are the thresholds beyond which belief systems become unrecoverable?* While the models in this book are computational, the stakes are human. As such, it advocates for a shift in how we approach infodemic management.

I offer this book to scholars, students, public health professionals, platform designers, and anyone invested in the future of truth and knowledge. As a cornerstone of knowledge mapping, it is written with the hope that by understanding how belief systems form and fracture, we can begin to rebuild shared meaning in a time that desperately needs it.

Bloemfontein, South Africa
July 2025

Dr. Herkulaas Combrink

Acknowledgments This book is the culmination of years of exploration, collaboration, and shared intellectual curiosity. It would not have been possible without the unwavering support, patience, and grounding presence of my partner, my children, and close friends, who reminded me, often gently, that life extends beyond simulations and belief systems. Your presence has been my compass.

I am deeply grateful to the officials within the Departments of Health, UNICEF, and the World Health Organization who generously shared their insights and institutional experience. Their commitment to advancing global health and digital trust provided a critical policy lens through which this work evolved.

Across the globe, I have been privileged to collaborate with exceptional colleagues whose intellectual contributions shaped this book in both visible and subtle ways. I would like to extend particular thanks to Prof. Sevias Guvuriro (University of the Free State), Prof. Rahul Katarya (Delhi Technological University), Prof. Alex Broadbent (Durham University), Prof. Benjamin Smart (University of Johannesburg), Prof. Benjamin Rosman (University of the Witwatersrand), and Prof. Vukosi Marivate (University of Pretoria). Your scholarship, mentorship, and dialogue pushed this work beyond disciplinary boundaries and into new theoretical terrain.

To my Ph.D. and Master's students, Phelokhazi Mkungeka, Valusha Oelofse your questions, code, frustrations, and breakthroughs were central to this journey. Thank you for the energy and rigour you brought to every line of code, every hypothesis, and every hard-earned insight. To my 2023, 2024, and 2025 honours students—your subtle frustrations and painful learning experiences humbled me every time of where I started this journey myself.

I am indebted to the contract staff and research collaborators who supported simulations, modelling, and data pipelines, often under immense technical pressure. Your dedication made the ambitious scope of this book possible.

Special thanks to the Research Network, Alber van Eck and the High-Performance Computing facilities at the University of the Free State. This work would quite literally not have run, figuratively or computationally, without your infrastructure and technical support.

Finally, I wish to thank the leadership of the University of the Free State for their belief in this vision and their steadfast support in building an inclusive, globally relevant ecosystem for computational social science and digital epistemology. Your backing helped transform a bold research idea into a completed work.

To all who contributed, named and unnamed, thank you for making this possible.

Competing Interests The author has no competing interests to declare that are relevant to the content of this manuscript.

Contents

Abbreviations

ABM	Agent Based Modelling
AI	Artificial Intelligence
COVID-19	Coronavirus Disease
ISB	Information Seeking Behaviour
ML	Machine Learning
RL	Reinforcement Learning
SA	Sentiment Analysis
SST	Social Stress Indicator
TM	Topic Modelling
WHO	World Health Organization

Chapter 1
Introduction to Computational Infodemiology

1.1 What is an Infodemic?

Can we truly try to understand the complex discourse that occurs in digital media? When misinformation spreads, can we map the belief systems behind the discourse? Are there indicators that exist that explain social stress or potential harm that may arise from misleading sentiments and disinformation? These are some of the fundamental questions that Computational Infodemiology tries to address through an interdisciplinary perspective. To understand infodemiology, let us unpack what an infodemic is.

The term "infodemic" was popularised by the World Health Organization (WHO) during the COVID-19 pandemic to describe the surge of information surrounding the crisis, even though the term was first described to outline the SARS and MERS related misinformation in 2001. Infodemics extend beyond health crises, encompassing political, environmental, and social domains. They thrive in environments characterised by high levels of uncertainty, often fuelled by the proliferation of social media and other digital communication platforms [118]. The sheer quantity of information, coupled with the speed of its dissemination, creates a cacophony of competing narratives. The digital revolution has facilitated instantaneous sharing of information through platforms like social media, messaging apps, and online forums. However, this rapid exchange often leads to information overload, making it difficult for individuals to discern credible sources [109]. The constant stream of updates, from breaking news to trending hashtags, can overwhelm even the most discerning audiences. An unintended consequence of an abundance of information among uncertainty is that the information seeking behaviour of people, often fuelled by instant gratification, generate echo chambers that reinforce misinformation. Sometimes these narratives are absurd, and at other times, they cause harm. In a study conducted by [78], misinformation related to COVID-19 in South Africa through digital media showed the level of vaccine related misinformation in this context, and coupled with the vaccine uptake over time, the low vaccine uptake of 2021 in

H. Combrink, *Computational Infodemiology*,
SpringerBriefs in Computer Science, https://doi.org/10.1007/978-3-032-10130-3_1

South Africa was quantified. Although there are multiple factors that play a part in the vaccine uptake challenges experienced in South Africa, misinformation, uncertainty, echo chambers, social norms, and reinforcement through constant feedback, rapid engagement, and a social reward loop that keep users engaged were a part of the problem.

Competing narratives emerge as individuals and organisations fight for attention, creating a chaotic information environment that is professional, social, and societal [22]. This phenomenon is exacerbated during crises, where the need for immediate answers collides with the scarcity of verified data. For instance, during the COVID-19 pandemic, the deluge of unverified claims and speculative theories confused public understanding and complicated decision-making [36]. The combination of a crisis, uncertainty, the right amount of online engagement, some truth, some opposing views, and a global audience is the perfect breeding ground for confirmation bias, warped social norms and perpetuate toxic online behaviour that contribute toward social stress that may cause public harm [40]. To understand this, misinformation is a key driver of the concept. Amidst an infodemic, three primary overarching ideas emerge to help us understand the broad classifications of information namely an information void, misinformation and disinformation [48].

Information voids are fertile ground for speculation and conjecture [116]. These voids arise when the demand for information exceeds the supply of credible content. For example, during the early days of the COVID-19 pandemic, the limited understanding of SARS-CoV-2 transmission created a knowledge vacuum that was quickly filled with unfounded theories about 5G technology causing the virus [22]. This gap enabled pseudoscientific explanations to gain traction across WhatsApp and YouTube, especially in regions where official communication was delayed or inconsistently translated. Information voids can also be observed in areas of new policy rollouts or scientific developments, such as artificial intelligence regulation or vaccine efficacy updates, where public curiosity is high, but expert consensus is still evolving [40].

Algorithmic systems often exacerbate voids by amplifying content that receives high engagement rather than factual accuracy. As Boyd and Golebiewski [44] explain, search engines and recommendation algorithms are not neutral, when users input vague queries, they may be led to extremist content or sensational conspiracy theories [116]. Thus, the concept of a void is not simply the absence of information, but the presence of attention misdirected toward low-quality or manipulative content.

Misinformation refers to inaccurate or misleading content shared without malicious intent. While often perceived as benign, misinformation can cause real-world harm by shaping beliefs and behaviours. One prominent example was the claim that consuming large amounts of garlic could prevent COVID-19 [116]. Despite the lack of biomedical evidence, this myth circulated widely, shared even by well-meaning individuals who believed it could protect others [51]. The spread of such content is influenced by emotional framing, cognitive heuristics, and the trust placed in interpersonal networks [92]. Unlike disinformation, misinformation thrives on credibility cues rather than manipulation, such as the appearance of authority or scientific terminology. In the South African context, vaccine misinformation often took the form of community-specific rumours, such as the claim that COVID-19 vaccines would alter DNA or were part of a depopulation agenda [78]. These narratives were not necessarily born out of malice but circulated within informal digital communities that lacked access to formal clarification.

Disinformation is the deliberate creation and dissemination of false or deceptive content intended to mislead, influence, or manipulate. Disinformation campaigns are often systematic and political in nature, leveraging coordinated networks of bots, trolls, or influencers. For instance, during the 2020 U.S. elections, state-sponsored disinformation campaigns used targeted memes and fabricated news stories to exacerbate racial tensions and erode trust in electoral processes [102]. In many cases, these campaigns weaponize existing social divisions, using disinformation as a tool of psychological and ideological warfare.

A localised example includes the disinformation surrounding xenophobic violence in South Africa, where manipulated videos and inflammatory posts falsely portrayed immigrants as perpetrators of crime, fuelling unrest and vigilantism [82]. These instances highlight the harmful intersection of disinformation with populist rhetoric and algorithmic visibility. Disinformation detection often relies on ensemble models combining natural language processing, bot detection algorithms, and provenance tracing to identify coordinated inauthentic behaviour.

While these categories of voids, misinformation, and disinformation are conceptually distinct, they often overlap. A single piece of content may originate in an information void, be amplified as misinformation by well-meaning users, and later be weaponised into disinformation by coordinated actors. Therefore, Computational Infodemiology models must be capable of not only categorising these types but also tracing their transitions and network effects. This requires high-resolution data, cross-platform surveillance, and a deep understanding of sociocultural context.

Misinformation often arises when uncertainty meets cognitive biases which leads to misinterpretations, or a lack of access to accurate information. For example, during health crises, well-meaning individuals may share unverified home remedies out of a genuine desire to help. Disinformation, on the other hand, is intentionally crafted to deceive, often for political, financial, or ideological gain. Disinformation campaigns use sophisticated strategies, including the use of bots, deepfakes, and fake news websites, to amplify false narratives. These tactics have been used widely in political contexts to sow discord, influence elections, and undermine trust in institutions like governments and academic universities. The interplay between misinformation and opportunity creates a feedback loop, where intentional falsehoods fuel the spread of unintentional inaccuracies, further complicating efforts to establish truth.

1.2 Algorithmic Feedback Loops and Contemporary Information Dynamics

In the evolving landscape of digital communication, the algorithmic feedback loop has become a central mechanism shaping contemporary information dynamics. This loop functions as a recursive system whereby user interactions such as clicks, shares, and viewing duration are recorded, processed, and used to tailor subsequent content exposure [15, 77]. The consequence is a form of behavioural reinforcement in which the content a user engages with most is algorithmically prioritised, thereby creating a self-reinforcing cycle of information exposure. Over time, this process leads to the construction of highly personalised and ideologically consistent information environments that are shaped not by epistemic diversity or factual integrity, but by engagement metrics optimised to retain user attention [21].

Unlike traditional mass media systems, which operate under editorial oversight and public accountability, contemporary algorithmic platforms function through opaque, dynamic mechanisms that privilege predictive modelling and real-time personalisation. The result is a shift from a shared public information space to a fragmented digital environment where individuals are increasingly exposed to content that affirms their prior beliefs [120]. This structural process is central to the formation of echo chambers, where exposure to dissenting viewpoints is minimised, and belief reinforcement becomes the norm [103].

The algorithmic feedback loop contributes to several critical socio-political effects. The most prevalent among these is the amplification of confirmation bias, a well-documented cognitive phenomenon whereby individuals preferentially seek, interpret, and remember information that supports their existing views. On social media platforms, confirmation bias is not only reproduced but algorithmically enhanced through engagement-based ranking systems that prioritise emotionally resonant or polarising content [92]. In this way, the feedback loop facilitates the rapid and disproportionate spread of misinformation and disinformation, particularly in high-uncertainty contexts such as pandemics, political crises, and environmental disasters [40, 51]. Furthermore, algorithmically reinforced exposure to homogenous information can deepen ideological divisions and undermine social cohesion. Group polarisation, whereby deliberation among like-minded individuals leads to more extreme positions, is frequently observed in digitally mediated discourse [103]. The feedback loop's influence on polarisation is further intensified by the socio-technical affordances of platforms that privilege high-arousal content and reward virality over veracity [112]. This creates conditions in which small-scale falsehoods can evolve into large-scale belief systems, sustained by social proof, repetition, and algorithmic recommendation [48]. While the algorithmic feedback loop can reinforce harmful information behaviours, it also has the potential to facilitate positive forms of mobilisation and knowledge dissemination. For instance, digital activism movements such as #MeToo and #BlackLivesMatter have demonstrated how algorithmic virality can be harnessed to amplify marginalised voices and catalyse social change [53]. However, this duality underscores the need for a nuanced understanding

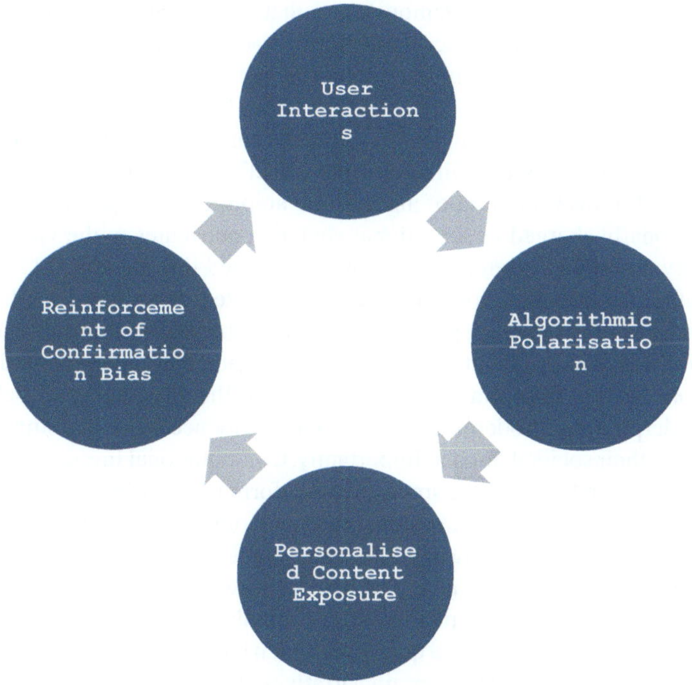

Fig. 1.1 Reinforcement feedback loop reinforcing misinformation

of digital algorithms not as inherently detrimental or beneficial, but as socio-technical systems whose effects are contingent upon design choices, user practices, and institutional governance [49]. As such, user interaction leads to algorithmic polarisation, which in turn validates the echo chambers that personalise misinformation leading to a reinforcement of confirmation bias within the ecosystem (Fig. 1.1).

At the heart of contemporary information dynamics lies a critical intersection between digital algorithms and societal polarisation. Contrary to the common misconception that algorithms are neutral, autonomous "black boxes," they are in fact socio-technical constructs embedded within institutional, economic, and cultural logics [77]. They are designed by humans, optimised through data-driven feedback loops, and operate within broader ecosystems of commercial imperatives and platform governance norms.

These algorithmic personalisation systems, while enhancing user engagement and platform retention, have the unintended effect of deepening ideological silos and reducing users' exposure to diverse or dissenting perspectives. This structural reinforcement of cognitive homogeneity leads to what [103] describes as "group polarisation," where like-minded individuals, exposed only to confirming narratives, become more extreme in their viewpoints over time. In the context of contentious public issues such as vaccination, immigration, or climate policy, such polarisation, undermines the possibility of deliberative democracy and cross-cutting dialogue

[120]. Empirical research has demonstrated that echo chambers not only reduce informational diversity but also accelerate the spread and acceptance of misinformation. Within such epistemic bubbles, content is more likely to be believed when it resonates with users' existing beliefs, regardless of its factual accuracy [92]. The repetition of misinformation in homogenous networks enhances its perceived credibility, a phenomenon known as the "illusory truth effect" thus further entrenching false beliefs [33]. Moreover, the role of algorithmic amplification, particularly in prioritising emotionally charged or sensational content, contributes to the viral spread of low-veracity narratives, often at the expense of verified information [112].

Computational Infodemiology emerges in this context as an interdisciplinary field that integrates natural language processing, network theory, behavioural science, and ethical governance to systematically monitor, simulate, and intervene in the spread of information. This discipline seeks to illuminate the structural and behavioural patterns underpinning infodemics and to provide evidence-based recommendations for mitigating their societal impact. Importantly, Computational Infodemiology seeks not only to detect and analyse the spread of misinformation (infoveillance) but also to simulate intervention strategies, evaluate their outcomes, and inform policy recommendations. By operationalising complex social signals such as emotional sentiment, engagement velocity, topic virality, and belief clustering this discipline provides a data-driven, ethically grounded methodology for making sense of digital discourse in real-time. It aims to answer critical questions which include "How do misinformation ecosystems evolve? What makes certain narratives more resilient or dangerous? How can public health, education, and governance systems be fortified against future infodemic threats?". Therefore, the study of algorithmic polarisation, echo chambers, and the societal consequences of information disorder is no longer optional it is essential for the future of democratic societies, public health, and global digital wellbeing.

1.3 Theoretical Fundamentals of Computational Infodemiology

Systemic intervention is a theoretical and methodological framework situated within the broader field of critical systems thinking, developed and advanced Midgley [73]. The framework is concerned with understanding and addressing complex, interconnected problems through reflexive and pluralistic systems practice. Rather than assuming a single 'correct' model or approach, systemic intervention emphasises the need for boundary critique, methodological flexibility, and the inclusion of multiple stakeholder perspectives in designing interventions in complex, uncertain, and value-laden contexts [73, 74]. Central to this framework is the idea that all decisions about how to understand and intervene in a system involve boundary judgements, choices about what is included, what is excluded, and whose voices are represented. Systemic intervention evolved in response to limitations within classical systems science, which often presupposed closed systems, fixed structures, and objective control. Midgley [73] situates systems thinking within a more politically and ethically reflexive paradigm, integrating critical systems heuristics [110] and soft systems methodology [19]. Systemic intervention promotes systemic pluralism by encouraging the integration of diverse methodologies, qualitative, quantitative, interpretive, and computational in tackling wicked problems. Intervention, in this context, is not understood as linear problem-solving, but as a continuous and reflexive process of negotiating complexity, making assumptions explicit, and adapting methods to context [121]. This makes systemic intervention particularly suited for ill-defined, dynamic problem domains with competing interests and high degrees of uncertainty. Another central contribution of systemic intervention theory is its commitment to ethical engagement. The framework critiques technocratic approaches that claim neutrality and instead call for ethical transparency, boundary critique, and constructive conflict among differing values and worldviews [75]. This makes the framework especially powerful in areas where knowledge is contested, institutional trust is fragile, and interventions must be context-sensitive and deliberative.

When applied to the context of digital and social media, systemic intervention offers a lens to interrogate the design, governance, and societal impact of online communication platforms and the recommendations and interventions that may be implemented to manage an infodemic, if one is present. Social media platforms are complex sociotechnical systems with multiple interacting subsystems algorithms, content creators, users, moderators, advertisers, and regulatory bodies each with their own goals, values, and power relations. These platforms are also emergent given their structure and outputs are not fully predictable due to feedback loops, adaptive user behaviours, and opaque algorithmic interventions [43, 111].

Systemic intervention helps uncover how platform designs make implicit boundary judgements such as whose content is promoted, whose voices are amplified or silenced, and what counts as credible or harmful information. For example, the prioritisation of engagement-based metrics in algorithmic design can be seen as an intervention that values virality over epistemic soundness [15].

Systemic intervention strength lies in its capacity to support tailored, participatory interventions that are reflexive to local conditions, especially when addressing deeply situated issues such as vaccine hesitancy within specific communities [73, 121]. For instance, consider the case of a health communication team seeking to reduce vaccine hesitancy in a geographically and culturally distinct urban neighbourhood. Here, systemic intervention facilitates a comprehensive mapping of the intervention's boundaries of relevance, ensuring that the design of public messaging strategies aligns with the contextual, sociocultural, and political conditions of that specific community [97, 110].

In practical terms, this entails an explicit boundary analysis that includes, but is not limited to the geographic scope of the campaign, the linguistic and cultural preferences of the community, the resource limitations (both human and financial), the historical relationship between public health institutions and the local population, the preferred social behaviour change communication strategies, and the presence of trusted gatekeepers or community leaders [27, 56]. Each of these dimensions represents a subsystem of the broader intervention ecology and must be examined as such. Rather than implementing a top-down or generic public health directive, systemic intervention encourages the co-construction of problem framings and solutions with the affected community [75]. This ensures that health communication is not only technically sound but epistemically and culturally resonant. A systemic intervention approach in this scenario might begin with stakeholder identification and engagement, particularly targeting individuals with local legitimacy such as religious leaders, traditional healers, local councillors, or school principals. Engaging these figures is not merely instrumental, it reflects an ethical stance towards inclusive systems practice, recognising that local narratives, fears, and lived experiences significantly shape health-related behaviours [37, 46]. Through this engagement, practitioners can surface dominant discourses surrounding vaccines, map areas of uncertainty, and understand the local epistemologies of trust and risk [122].

Based on these insights, multi-disciplinary teams comprising health promoters, public health officials, social scientists, linguists, and digital communication specialists can collaboratively design localised health messaging campaigns in vernacular languages. These messages may be adapted to circulate not only through traditional media (e.g., radio or pamphlets) but also via social media platforms used by the community, thus bridging online and offline information ecosystems [64]. Importantly, systemic intervention allows these messages to be dynamically adjusted based on community feedback, real-time misinformation monitoring, and the evolving nature of vaccine concerns.

Critically, systemic intervention provides an analytical structure to articulate and monitor boundaries identifying what is included, excluded, or marginalised in the design and implementation of interventions [73]. In this example, boundaries might include national health policy constraints, linguistic diversity (such as different dialects of local languages), the available communication channels (physical and on digital media), infrastructural limitations, and the evidence base for different behavioural strategies [88]. By making these boundaries explicit and open to critique, systemic intervention supports adaptive learning and ethical deliberation, ensuring that interventions do not simply disseminate information but rather engage meaningfully with the community to build trust, address hesitancy, and support collective health decision-making [16].

In Computational Infodemiology, this translates into the integration of computational tools (e.g., natural language processing, network analysis, anomaly detection) with qualitative and participatory methods, such as digital ethnography and stakeholder mapping. This hybrid approach draws from multiple traditions to enrich systems practice and avoid methodological hegemony [73, 97]. For example, one might use data-driven models to map misinformation networks while also using stakeholder interviews to understand the motivations and vulnerabilities of affected communities. Misinformation in digital environments is not merely a content problem but a systemic problem, rooted in epistemic fragmentation, institutional distrust, algorithmic opacity, and sociopolitical polarisation. Addressing it requires moving beyond linear solutions (e.g., fact-checking alone) toward systemic strategies that engage with the broader ecology in which misinformation thrives [30]. This is where systemic intervention proves essential.

Under conditions of uncertainty such as health crises, political upheaval, or environmental disasters, traditional top-down information dissemination often fails to gain traction. People seek alternative narratives that resonate with their lived experiences or identities. Systemic intervention theory reminds us that any attempt to counter misinformation must first understand the boundaries of how affected populations define knowledge, trust, and authority. For instance, systemic boundary critique can help elucidate why certain communities reject public health messaging: not because they are 'irrational', but because their knowledge boundaries differ from those assumed by formal institutions [55].

Furthermore, systemic intervention foregrounds reflexivity and ethical plurality, both of which are crucial in responding to infodemics. Interventions that fail to acknowledge plural values, power asymmetries, or local epistemologies risk perpetuating harm or reinforcing existing inequalities. Computational Infodemiology, informed by systemic intervention, can thus move from a reactive stance (detecting and removing harmful content) to a proactive and inclusive strategy that simulates interventions, co-designs solutions with stakeholders, and ethically evaluates the consequences of different courses of action. In essence, systemic intervention offers a theoretical scaffold for designing socially and ethically responsive computational approaches to misinformation. Building upon the reflexive and participatory insights of systemic intervention, the next theoretical foundation for Computational Infodemiology concerns the mechanisms through which social norms are formed, stabilised, and contested in digital environments.

1.3.1 Systemic Intervention and Social Norm Formation

While systemic intervention equips us to understand the boundary dynamics and stakeholder perspectives within a complex system, social norm theory provides the behavioural and sociological scaffolding needed to examine how collective expectations around truth, trust, and knowledge emerge and evolve, particularly in networked digital spaces saturated with uncertainty.

Social norms are the informal, often unspoken rules that govern acceptable behaviour within a group or society. Rooted in social psychology and behavioural economics, the study of norm formation seeks to explain how collective expectations are established, maintained, and changed through interpersonal interactions and institutional reinforcement [10, 123]. A widely accepted model of norm formation is the three-stage model, which conceptualises norm development as a sequential process including emergence, enforcement, and internalisation [66, 106]. Emergence refers to the initial articulation or disruption of behavioural expectations within a community. This can occur organically or be introduced through external interventions. Enforcement involves the use of social sanctions or rewards to reinforce norm adherence. This can be formal (e.g., platform moderation) or informal (e.g., peer approval). Internalisation occurs when individuals adopt the norm as a personal standard, even in the absence of social pressure.

This model draws from the tradition of norm cascades in international relations theory [34], social influence theory in psychology [20], and more recent behavioural insights about descriptive and injunctive norms [11]. Importantly, norms are not static they are dynamically co-produced through ongoing social interaction and are sensitive to context, authority cues, visibility, and perceived consensus. What complicates digital media and specifically social media, is that the influence can come from anywhere in the world that has access to the conversation. In other words, in public online engagements on social media platforms, the participants of the discourse are important, not where they are geographically situated. This poses a problem as influence may come from anywhere in the world, making it difficult to intervene physically (through for example a health campaign). This means that knowing the specific stage an echo chamber or societal topic is, is an important task because the interventions will differ between knowing if the echo chamber or public conversation is at the emergence, enforcement, and/or internalisation phase.

The theory also recognises the role of pluralistic ignorance, where individuals misperceive the prevalence of a belief or behaviour among their peers and thus refrain from challenging it, even if they personally disagree [93]. This misalignment between private beliefs and public behaviour can stabilise false or harmful norms, an insight especially relevant in the context of misinformation and infodemics. In digital environments, social norm formation operates through highly visible, algorithmically mediated interactions. Likes, shares, hashtags, comments, and follows all function as digital cues that signal collective endorsement or disapproval. These cues facilitate norm emergence and enforcement by making behaviours and opinions observable to others [17]. For instance, when users repeatedly see antivaccine narratives receiving high engagement, they may come to perceive these views as normative, even if they privately disagree, thus lowering the threshold for sharing similar content themselves.

The three-stage model can be mapped directly onto digital platforms. Emergence often occurs when a message or behaviour (e.g., misinformation, social justice advocacy) goes viral, disrupting existing discursive norms. Enforcement happens through social feedback loops such as comment reinforcement, cancellation, blocking, or endorsement by influencers and verified accounts. Internalisation follows when repeated exposure and social proof lead users to adopt these norms as part of their digital identity or group affiliation. Importantly, social media platforms do not merely reflect social norms, they shape them. The visibility and amplification mechanisms of algorithms may accelerate the norm formation process by rapidly exposing users to majority-like behaviours and dominant narratives, especially when the discourse is laden with a certain degree of uncertainty [118]. Computational Infodemiology can apply this model to trace how misinformation gains normative traction. For example, a fringe conspiracy theory may first emerge in an obscure online forum, then be picked up by influencers (emergence), receive widespread engagement (enforcement), and ultimately become a common belief within a certain community (internalisation). Tools such as sentiment analysis, network clustering, and meme tracking can empirically measure these norm dynamics in real time [21, 29]. In contexts of uncertainty, such as pandemics, elections, or natural disasters, social norm formation becomes an especially potent mechanism through which misinformation can spread and become entrenched.

When authoritative information is delayed, contradictory, or inaccessible, people look to peers and influencers for behavioural cues. This makes norm emergence highly susceptible to informational cascades, where early adopters of false claims influence the wider population, regardless of empirical accuracy [13]. The internalisation of misinformation as normative belief, particularly when reinforced by community identity, distrust of institutions, or perceived persecution, poses significant challenges to public health and governance. For example, during the COVID-19 pandemic, communities in which vaccine scepticism became a social norm saw dramatically lower uptake, even when accurate information was made available [64]. In such cases, attempts to correct misinformation through fact-checking may be insufficient or counterproductive, especially if they fail to address the underlying social norms and trust structures. It is therefore vital that a proxy for the stage of social norm formation be tracked and measured so that the interventions and strategies can be tailored to that specific problem, both based on the boundaries identified within the system to produce a meaningful intervention and more importantly, which boundaries fit which stage of the social norms within groups and individuals. A "norm-sensitive infodemiology" approach, therefore, requires more than content moderation or debunking. It involves identifying norm influencers, understanding the symbolic function of misinformation within communities, and designing norm-shifting interventions. These may include public health campaigns that leverage injunctive norms (what people ought to do), showcase diverse community narratives, or use credible messengers to model behaviour change [11, 106]. Ultimately, integrating social norm theory into Computational Infodemiology enables a more relational and dynamic understanding of how false beliefs evolve, not simply as isolated cognitive errors, but as socially embedded expectations that are shaped, enforced, and resisted within complex digital societies. Having examined how systemic intervention offers a methodological structure for contextual boundary analysis, and how social norm formation explains the behavioural dynamics underpinning misinformation, the final theoretical pillar of Computational Infodemiology turns to the moral and political dimensions of human agency. To address not only how misinformation spreads but why it matters who influences the people (or bots) spreading the information, we draw from a capabilities approach, particularly spheres of influence, a concept grounded in the normative idea of what individuals must be able to do and be in order to live a life of dignity and justice and where these concepts are drawn from [87].

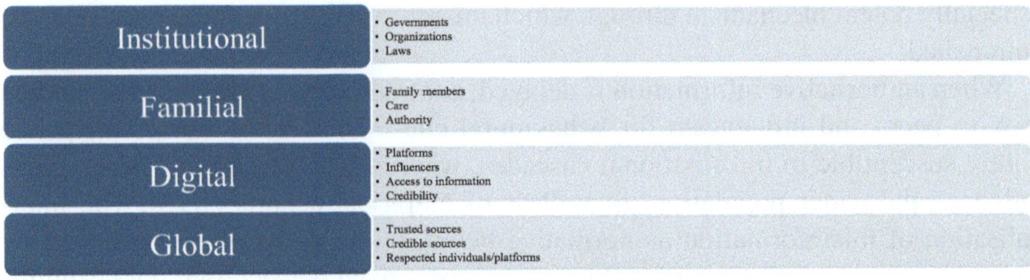

Fig. 1.2 Illustration of spheres of influence

1.3.2 Spheres of Influence and Misinformation

Capabilities Approach is a philosophical and political framework that reorients social justice around what individuals are substantively free to achieve. At the core of Nussbaum's political philosophy is the recognition that individuals do not operate in isolation but exist within overlapping domains of influence, familial, social, institutional, economic, digital, and transnational [87]. These spheres of influence describe the relational environments in which power is exercised, and decisions are shaped, often beyond formal governance structures. Each sphere entails dynamics of authority, care, communication, and responsibility, influencing not just individual outcomes but also collective norms and social priorities (Fig. 1.2).

Within this framework, influence is not uniformly distributed. Individuals and institutions vary in their ability to affect others, depending on their social position, perceived legitimacy, and communicative reach. For instance, a well-respected community elder may exert significant influence over local norms and decisions, more so than formal authorities or external actors. Similarly, religious leaders, schoolteachers, or informal digital influencers may play pivotal roles in shaping community interpretations of health guidance, political developments, or scientific claims. These power differentials are context-specific, temporally dynamic, and often underpinned by cultural, linguistic, or historical legitimacy [99, 122].

Spheres of influence thus offer a conceptual lens through which we can understand how informational authority is distributed, how trust is constructed, and how individuals and groups become vulnerable or resilient to false or misleading narratives. Applied to digital media, spheres of influence help explain how individuals' engagement with information is shaped not solely by content, but by the relational and technological systems through which that content is accessed and validated. Digital platforms, especially those structured around algorithmic curation and social amplification, constitute powerful spheres of influence in themselves. They not only determine what content is visible but also signal what is considered popular, credible, or worthy of further attention [15, 43]. In these environments, influence is both platform-generated (through algorithmic amplification, trending lists, and verified accounts) and user-generated (through followers, engagement metrics, and reputation). The visibility of actors, such as influencers, celebrities, or pseudoscientific personalities, can produce the illusion of legitimacy, especially when combined with viral sharing patterns and community endorsement. This amplifies the role of informal epistemic authorities in shaping belief systems, even when these figures lack scientific or institutional expertise [21]. Moreover, the sphere of digital influence intersects with offline hierarchies. In many contexts, respected community figures are active both online and offline, acting as conduits through which information, values, and misinformation flow. For example, a trusted local pastor with a strong Facebook presence may significantly shape vaccine attitudes within a congregation. The same applies to political organisers, social movement leaders, or youth micro-celebrities on social media platforms [83]. Understanding who influences whom, and in what sphere, becomes critical for effective public engagement and infodemic response. In Computational Infodemiology, this theory prompts a network-aware approach in identifying the overlapping spheres, digital, social, cultural, that mediate access to information, and mapping the actors and nodes that carry disproportionate influence within these domains.

In moments of high uncertainty, such as health crises, political unrest, or climate disasters, these spheres of influence become decisive in shaping public understanding and response. When institutional information is inconsistent, delayed, or distrusted, individuals turn to proximate and familiar sources of authority within their spheres of influence. These may include community leaders, peers, religious figures, or digital creators whose influence stems from cultural proximity or emotional resonance rather than formal expertise [55]. This shift has direct implications for the spread and containment of misinformation. If false claims originate from, or are endorsed within, trusted spheres of influence, they are more likely to be accepted and less likely to be questioned. Conversely, corrective information delivered through unfamiliar or distrusted sources, even if factually accurate, may be ignored or actively resisted. Therefore, understanding and intervening within spheres of influence is key to shaping informational resilience. In practical terms, this means that infodemiological interventions must be relationally and contextually informed. They must identify which actors hold influence, in which communities, and on which platforms. Interventions should aim to work with these actors, not against them, to design communication strategies that are linguistically accessible, culturally resonant, and ethically attuned to local needs and values. Contextually this becomes challenging when considering where in social norm formation the specific individual or groups are, and more importantly, the boundaries that persist within the context of the systems and the interventions that may be enacted on within these communities, whether it is on a digital platform or in person. By focusing on these theories, Computational Infodemiology moves beyond a one-size-fits-all model of misinformation correction. It encourages a pluralistic, systems-informed, and ethically reflexive approach to digital intervention, recognising that what people come to believe is shaped as much by who tells them, how they are told, and in what sphere, as by the content itself. Given this support, mapping these infodemics, beyond the words used in the discourse, become important. To do so, a framework is needed to map these concepts computationally, and in Chap. 2, this will be addressed.

Chapter 2
Agent Based Modelling and Reinforcement Learning in Computational Infodemiology

2.1 Background

Misinformation is a critical challenge in information dissemination and content management, especially in digital information ecosystems, with severe consequences if not approached properly. The ability to model misinformation spread, or being able to model an infodemic using computational approaches is at the core of Computational Infodemiology, but to conceptualise exactly how to model an infodemic, is a difficult task. Traditional methods, such as rule-based fact-checking and reactive content moderation, often fail to scale effectively in dynamic online environments. Just imagine the manpower required to categorise online content if it was produced at a rate of 10,000 microblogs and posts per second? Part of the problem that traditional monitoring and evaluation metrics have is that it does not account for the exploration and random chance that human beings introduce into a system. Another challenge is that there are certain complexities that cannot be captured by traditional statistical models alone if we work with dynamic systems that represent online interactions and online discourse. To address this level of uncertainty, I look toward agent-based modelling as a viable framework for quantitatively capturing some of the more nuanced behavioural variations that traditional statistical models may not fully account for. This chapter will outline some of the foundational ideas, approaches and methods associated with agent-based modelling for Computational Infodemiology, so that we can start thinking about modelling such an infodemic.

The prediction of information spread remains a complex challenge, primarily due to its entanglement with qualitative factors such as social influence, normative behaviour, and emergent network dynamics [63, 114]. As previously discussed in Chap. 1, modelling infodemics as complex systems introduces multifaceted behavioural and environmental considerations, often shaped by cognitive biases and sociotechnical interactions. While acknowledging the intricate, nonlinear nature of information propagation is important, an overemphasis on qualitative nuance can

H. Combrink, *Computational Infodemiology*,
SpringerBriefs in Computer Science, https://doi.org/10.1007/978-3-032-10130-3_2

hinder scalable analysis and intervention design [61]. To overcome these limitations, computational approaches such as agent-based modelling (ABM) have become increasingly prominent in the study complexity [18]. ABMs simulate interactions among autonomous agents, each governed by context-specific behavioural rules, within artificial social environments. This bottom-up modelling technique allows for the capture of emergent patterns, the inclusion of heterogeneous agent behaviours, and the quantification of evolving social patterns, as seen in the fields of behavioural and experimental economics [45]. A key advantage of ABMs lies in their sensitivity to micro-level changes, which can cascade into large-scale shifts, mimicking real-world phenomena where small perturbations in individual behaviour or network structure produce disproportionate impacts [102]. Additionally, these models can incorporate multimodal data sources, such as social media analytics, longitudinal survey data, and behavioural logs, to enhance their realism and predictive power [25]. Beyond prediction, ABMs provide a controlled framework to simulate and test interventions, ranging from fact-checking mechanisms to policy nudges, enabling researchers to evaluate both intended effects and unintended consequences. For AGMs to effectively work, the correct environment and framework is needed.

2.2 An Introduction to Agent Based Modelling

Agent-based modelling (ABM) is a computational technique used to explore complex systems by simulating them as collections of independent units known as agents, each defined by unique attributes and behavioural protocols [70]. Unlike conventional modelling approaches that apply aggregate-level equations or average behaviours, ABM emphasises the bottom-up processes driven by individual-level interactions [18]. These agents, whether they represent people, animals, institutions, or abstract constructs, function autonomously and adapt their actions based on internal decision-making rules and their surrounding environment [1]. Building an agent-based model involves specifying both the characteristics of the agents and the logic guiding their behaviours. These behavioural rules may be simple, such as reacting to nearby agents, or more complex, involving conditional logic based on environmental inputs [18]. Once defined, agents operate within a simulated space, where they interact at each step of the simulation. These steps unfold over time, with agents continuously updating their status in response to internal states and external changes [95]. As an example, consider a ten-second simulation designed to track disease transmission through a network. The model uses a scale-free network, where a few nodes (hubs) have many connections, while most have relatively few [2]. In this setup, agents start in a neutral (grey) state and become infected (red) based on specified contact and contagion rules. Infection timing depends on both the agent's susceptibility (linked to node size) and its proximity to already infected agents. Smaller, more vulnerable nodes are infected more quickly than larger, more resistant ones, for instance, large nodes may infect small neighbours in two seconds, medium-sized ones in three

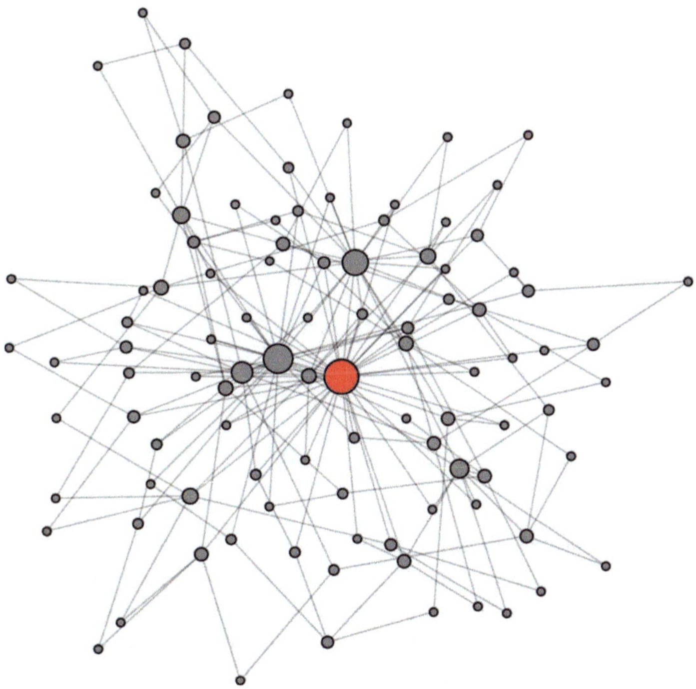

Fig. 2.1 First state of modelling disease spread within a room

seconds, and other large nodes in four. Only adjacent nodes can spread the infection, reinforcing the model's spatial realism. To introduce randomness, the simulation includes probabilities: a 50% chance per second that a large node will begin infecting others, compared to a 20% chance for smaller ones. These probabilistic elements help reflect the unpredictability found in real-world contagion scenarios [18]. Figure 2.1 depicts the model's initial state, from which the simulated contagion begins.

Under the defined transmission rules, low-degree nodes turn "infected" (i.e., red) more quickly upon exposure, yet because they have a lower overall chance of contracting the infection, by the ten-second mark the number of infected nodes among both small and large degree classes ends up being roughly equal. We can contrast this with a simulation that omits the degree-dependent susceptibility rule altogether. In doing so, it becomes immediately apparent that simply including or excluding this probabilistic vulnerability criterion yields markedly different infection patterns across the same network structure (Fig. 2.2).

In the simulation, the infection patterns in the rule-governed network (Fig. 2.2a) appear markedly different from those in the unconstrained counterpart (Fig. 2.2b). What makes these structures particularly valuable is their predictable, rule-driven temporal evolution, by recording node-state transitions at each second, I can apply time-series techniques to quantify how the contagion unfolds. Figure 2.3 therefore charts, side by side, the progression of "infected" (red) and "susceptible" (grey)

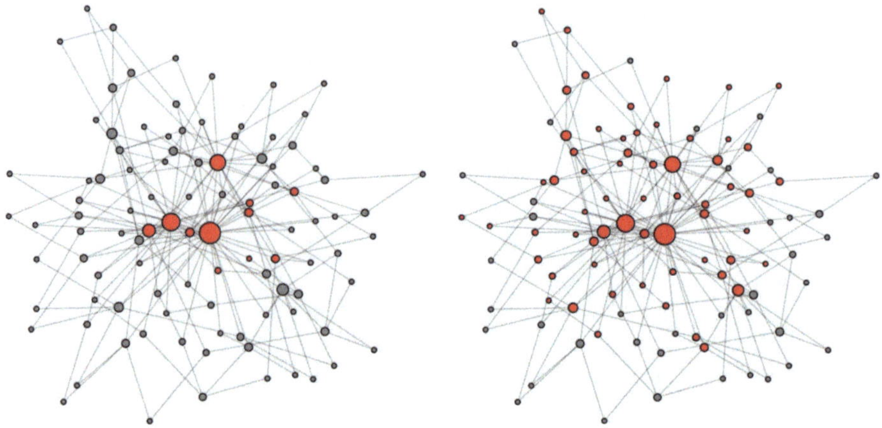

Fig. 2.2 Rule based network and unconstrained network

nodes in both the rule-based and non-rule-based models over time. Extending this framework to include recovery, we introduce a third state, "recovered" nodes, shown in green, to represent individuals who can no longer be reinfected. When this additional rule is incorporated and the network is simulated over a ten-second interval, the resulting trajectory of red, grey, and green nodes produces the pattern illustrated in Fig. 2.3.

When the model is executed, macroscopic regularities begin to surface from the multitudes of localised exchanges, regularities that are not readily inferred by inspecting any single agent's decision-rule in isolation. Take, for instance, a simulation of how an "infection" propagates through a network with each synthetic individual having a transmission potential based on factors such as infection rates, the proportion of its neighbours who have already transmitted the infection, or even its own predisposition to be infected. Although each agent follows relatively simple heuristics, the collective outcome over successive time steps can vary dramatically. In this instance, the infection may spread within isolated pockets of the network, cascade explosively once it surpasses a critical mass of well-connected nodes or large, smaller clusters, or propagate at a steady, moderate pace depending on the initial distribution of connectivity within the network. This phenomenon, known as emergent behaviour, occurs when system-level patterns, such as tipping points, threshold effects, or clustering dynamics, arise organically from individual-level interactions rather than from any centralised orchestration. In this way, ABMs affords us a lens through which to discern the "big picture" of a complex adaptive system by meticulously tracking the "small picture" of local rules and interactions. As such, it is especially well suited to investigating domains, like disease spread and, in this context, information diffusion, market dynamics, or epidemiology, where global phenomena are deeply contingent on the interplay of numerous, heterogeneous actors operating under simple, localised decision protocols. To understand the shape of the network for modelling these

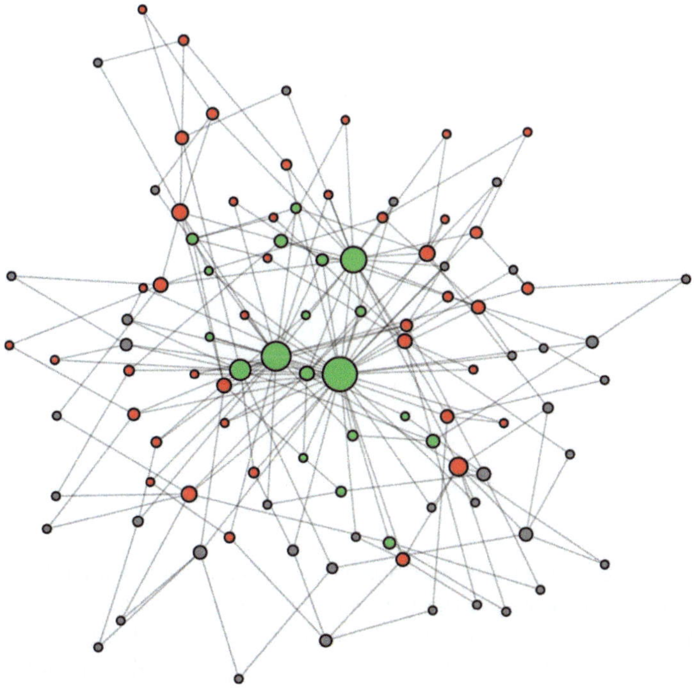

Fig. 2.3 10 s disease spread simulation

scenarios, it is important to consider different network topologies that can be used to represent ABMs.

2.3 Network Topologies

In ABMs, different network structures create varying patterns of information flow, contagion dynamics, and emergent behaviours (Fig. 2.5).

A random (Erdős–Rényi) network is constructed by linking each pair of agents with a uniform probability, resulting in a binomial (approximated by Poisson) degree distribution and no prominent hubs. Such graphs exhibit low clustering and short average path lengths, which promotes rapid but uniform diffusion of information across the network [2, 32]. The Watts–Strogatz small-world model begins with a regular lattice and then randomly rewires a fraction of its edges. This process preserves high local clustering while dramatically reducing characteristic path lengths, making the topology well-suited for phenomena requiring both cohesive local groups and efficient long-range communication, such as the spread of ideas or pathogens [114]. Scale-free networks emerge when new nodes preferentially attach to well-connected incumbents, producing a power-law degree distribution in which a

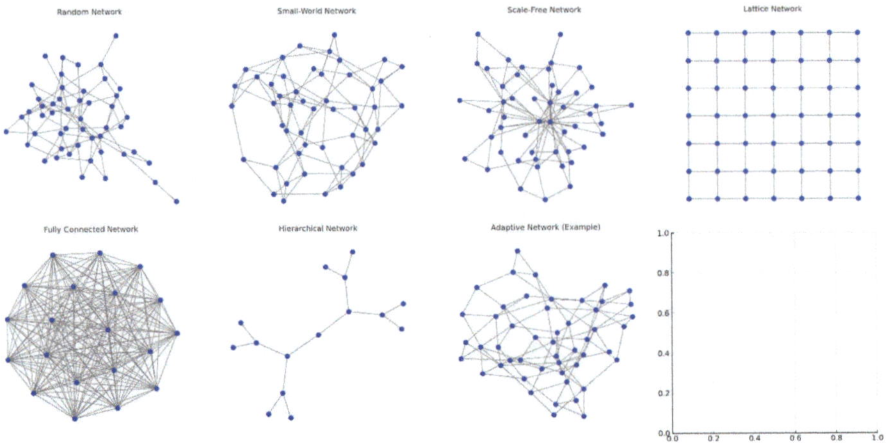

Fig. 2.5 Different network topologies

few hubs dominate connectivity, and most nodes remain sparsely linked. This heterogeneity confers robustness against random failures but acute vulnerability to targeted attacks, and it accurately captures the dynamics of systems where influential actors drive global processes, from viral content diffusion to financial contagion [2, 7]. In a lattice network, agents occupy positions on a regular grid (often toroidal), connecting exclusively to immediate neighbours. This arrangement yields uniform degrees and high clustering but long path lengths, reflecting scenarios where interaction is strictly local and information or influence propagates slowly, as in spatial ecological models [59]. A fully connected (complete) graph links every pair of nodes, minimising path lengths to one step and ensuring homogeneous exposure across the population. While this maximises diffusion speed and equitability of influence, the quadratic growth in edges renders it impractical for large systems [52]. Hierarchical networks exhibit a tree-like, multi-level organisation in which nodes are stratified into levels. Information flows are efficient along hierarchical chains, and roles are clearly delineated, but the failure of higher-level nodes can sever large subtrees, mirroring vulnerabilities in organisational or supply-chain structures [96]. Adaptive networks allow the topology itself to co-evolve with node states or behaviours, enabling links to form or dissolve based on interactions. This dynamic rewiring fosters emergent clustering and shifting connectivity patterns, making adaptive networks a powerful framework for modelling social learning, opinion dynamics, and cooperative behaviour in changing environments [47]. Each topology offers distinct advantages depending on the modelling objectives: random graphs for homogeneous diffusion, small-world for balancing locality and reach; scale-free for capturing hub-driven phenomena; lattice for purely local interactions; complete graphs for maximal connectivity; hierarchical structures for role-based flows; and adaptive networks for co-evolutionary dynamics. Table 2.1 below summarises the key characteristics of these different network types.

Table 2.1 Summary of network topologies of ABMs

Network type	Degree distribution	Clustering	Path length	Key properties
Random	Poisson	Low	Short	Homogeneous, no hubs
Small-world	Binomial/power-law	High	Short	Clustering with shortcuts
Scale-free	Power-law	Low	Short	Hubs dominate connections
Lattice	Uniform	High	Long	Localised interactions
Fully connected	Uniform	High	Very short	Global knowledge, fast diffusion
Hierarchical	Tree-like	Medium	Medium	Centralised control, vulnerability to failures
Adaptive	Dynamic	Variable	Variable	Evolves over time

As the field of Computational Infodemiology matures, each topology may illuminate different facets of information outbreaks. Mislove et al. [76] demonstrated that across Flickr, YouTube, LiveJournal, and other services, user connectivity adheres closely to a power-law, confirming that a handful of highly connected accounts drive most interactions. Likewise, Kwak et al. [62] found that Twitter's follower graph displays scale-free characteristics, with top accounts functioning as critical conduits for news dissemination and opinion formation. Lerman and Ghosh [67] extended these insights by examining how content on Digg and Twitter diffused as they showed that story popularity and spread velocity are both strongly modulated by the engagement patterns of high-degree nodes, underscoring the pivotal role of hubs in catalysing viral cascades. By embedding synthetic agents within a Barabási–Albert network, we capture not only the rapid, hub-mediated propagation of both credible information and misinformation, but we can also model it. This type of network's inherent heterogeneity resembles features that simpler random or lattice models cannot reproduce. This reliability of these networks to migrate instances across the network at different rates is crucial when simulating counter-infodemic interventions, but before this can happen, the fundamental modelling of information spread or the modus operandi behind them needs to be mapped. Moreover, because scale-free networks exhibit vanishing epidemic thresholds in the infinite-size limit, even low-virality memes can achieve widespread penetration if they engage key influencers, a phenomenon borne out in real-world social media studies. Consequently, scale-free topologies provide a realistic scaffold for modelling the emergence, acceleration, and mitigation of abstract concepts like agenda, sentiment, and belief system, making them the ideal foundation for the computational explorations that follow. For the remainder of this book, I will concentrate on scale-free networks as a representation of social media dynamics in the context of echo chambers. What will not be discussed in this book are the algorithmic biases created by advanced recommender systems nor the algorithmic bias generated by AI trained machine learning processes that classify content as credible or not based on the training data. The next steps are to map how to model agents that explore and more importantly, learn, as a proxy for different interactions within the network. For this, I will use a reinforcement learning (RL) framework.

2.4 Introduction to Reinforcement Learning in ABM

RL provides a field of study that creates a framework where a digital autonomous learner, referred to as an agent, uncovers effective decision strategies by interacting repeatedly with its environment and receiving only scalar feedback rather than explicit "correct" answers [113]. Over time, the agent learns from the feedback which action is best for the task it is trying to complete, and as such simulates how the agent interacts with an environment. In the process, the agent learns which choices yield a better cumulative reward, and what is learned is stored for later use. This means that RL agents interact with the environment in which they are, making decisions over the type of actions they can take, and explore different parts of the environment. This also means that the agent is required to have autonomy over the actions it may take. The action the RL agent makes on the environment is thus up to what the agent had learned from its interactions, rewards, framework, and defined environment. At the core of RL lies a Markov Decision Process (MDP), which frames the problem the agent is trying to solve in terms of environment snapshots (states), the agents possible moves (actions), which are probabilistic rules that govern how actions transition the system between states, a reward function that quantifies the immediate benefit or cost of each state–action pair, and a discount factor that balances the value of present versus future payoffs [94]. Unlike supervised learning, where correct outputs are provided a priori, RL agents must explore unknown state–action spaces, using reward outcomes to iteratively refine value estimates. It is this interaction of acting, getting feedback, and then summating a reward that enables the agent to learn. An RL experiment is akin to Pavlov's classical conditioning but frames the experiment as an MDP. In the Pavlov experiment, the ringing of a bell defines a state, salivation and withholding salivation are actions, and the subsequent delivery of food provides a positive reward signal. Through repeated pairings, the animal's "policy" (or learned context) converges on salivating in response to the bell alone, irrespective of whether food was presented [91]. Just like the Pavlov experiment, RL enables autonomous agents to acquire optimal decision policies through repeated interactions with an environment, receiving feedback signals rather than explicit supervisory labels [113]. RL algorithms strive to maximise the expected sum of discounted returns, to yield a higher reward over time [94].

There exist several strategies to maximise the cumulative return, given the degree of exploration and exploitation. For example, in ε-greedy (Epsilon-greedy) strategies an agent samples a random action and with probability, it exploits current value estimates and provide a simple yet effective heuristic. If the agent explores a new action, and the reward is less than the previous action, the agent reverts to the previous action and repeats the action. Other more advanced approach like the Upper Confidence

Bound (UCB) algorithm, allocates exploration proportional to each action's uncertainty estimate [4]. Several disciplines have used RL to map a more "human like" or biological process within people. One such example is computational psychiatry.

Computational psychiatry is a rapidly developing interdisciplinary field that seeks to understand mental disorders through the lens of computational models derived from neuroscience, psychology, artificial intelligence (RL), and statistical learning. Computational psychiatry provides a formal framework for interpreting how dysfunctional cognitive and neural processes can give rise to psychiatric symptoms. Rather than relying solely on descriptive diagnostics, computational psychiatry aims to construct mechanistic models that simulate perception, belief updating, reward learning, and decision-making in both healthy and disordered minds [39, 80]. These models enable researchers to infer underlying processes from observable behaviours and neurobiological signals, thus offering a new paradigm for studying psychiatric phenomena such as hallucinations, delusions, compulsions, and affective disturbances [105]. Within this field RL has become a particularly influential approach due to its ability to mirror how humans adapt behaviour based on environmental feedback. This structure closely aligns with how human and animal behaviour adjusts in response to rewards and punishments, particularly in the context of dopaminergic signalling pathways in the brain [98]. In computational psychiatry, RL models are used to simulate how reward prediction errors, which is the difference between expected and actual outcomes, influence learning and behaviour. These errors are believed to be encoded by dopaminergic neurons in the midbrain and are critical to updating beliefs about action-outcome contingencies. Deviations in this process, based on higher errors, have been implicated in a range of psychiatric disorders. For instance, individuals with major depressive disorder often exhibit reduced sensitivity to positive outcomes, which manifests as lower reward prediction error signals and blunted behavioural responses to reward [50]. Similarly, schizophrenia has been associated with deviations in prediction error signalling, leading to the inappropriate updating of beliefs and contributing to the formation of delusions [26]. In obsessive–compulsive disorder and anxiety disorders, RL models have been used to capture exaggerated threat prediction and overreliance on avoidance strategies, while ADHD has been associated with an impaired ability to balance exploration and exploitation during decision-making [36]. Importantly, RL models not only simulate behaviour but also yield parameters. Some of these include learning rates, discount factors, and exploration strategies which can be quantitatively linked to neural and clinical data. This capacity to bridge behaviour, brain function, and clinical symptoms is one of the key strengths of computational psychiatry [42]. By modelling both goal-directed (model-based) and habitual (model-free) decision systems, RL provides insight into the imbalance between cognitive control and automatic behaviour, a feature relevant in many psychiatric conditions [28, 124]. RL thus contributes to computational psychiatry by offering an algorithmic framework that closely mimics core human cognitive processes. It captures the dynamics of trial-and-error learning, reward sensitivity, and value-based decision-making [39, 105]. Another discipline where RL is being used to map social human behaviour is in behavioural and experimental economics.

In behavioural and experimental economics, insights are integrated from psychology, cognitive science, and economics to explain why human choices often deviate from the predictions of classical rational choice models. In behavioural and experimental economics, RL is used to model how people learn from experience in environments where outcomes are probabilistic, delayed, or influenced by others' actions. Rather than assuming perfect information and rational foresight, RL frameworks such as Q-learning or temporal difference learning allow for modelling bounded rationality. Bounded rationality relies on agents to update value estimates incrementally based on prediction errors. For instance, individuals repeatedly choosing between risky financial investments or consumer products often learn through trial and error, updating their expectations about returns or satisfaction over time [104]. A prominent application of RL in behavioural and experimental economics is in intertemporal choice. Intertemporal choice is decisions involving trade-offs between immediate and delayed rewards. Empirical research has shown that humans often exhibit temporal discounting, favouring smaller, sooner rewards over larger, delayed ones [14]. RL models with discount factors (γ) directly capture this tendency and allow researchers to examine how different populations exhibit different sensitivities to social distancing, reward delay, and incentives [57]. Moreover, RL-based models can simulate hyperbolic discounting, a pattern often observed in human behaviour that deviates from the exponential discounting assumed in classical economic theory [14]. In experimental economics and game theory, individuals update strategies based on the observed outcomes of others, as in the case of public goods games, auctions, or coordination tasks. RL frameworks such as multi-agent RL or Bayesian reinforcement learning have been used to simulate how individuals adapt behaviour in social contexts, learning not only from personal outcomes but also from social feedback, norms, and peer behaviour [12]. This allows researchers to map how behaviours such as cooperation, trust, or competition evolve in populations over time. RL provides a flexible, data-driven framework that better reflects how people learn and adapt over time, offering a powerful alternative to static decision models. Its integration into behavioural and experimental economic research has not only deepened theoretical understanding but also enhanced the design of policies and interventions aimed at improving decision quality and well-being. This makes RL for Computational Infodemiology ideal as an agent decision making framework within ABMs because the agents can "behave" more "human-like" by exploring information, making decisions and choosing what other agents to influence. Moreover, instead of focussing on a single agent, the focus is rather on how agents interact with one another, like on social media.

However advanced these algorithms are, they are not without flaws. RL agents often learn opaque policies, complicating interpretability in safety–critical applications. Another issue is that efforts to extract symbolic representations or causal structures from learned networks remain a challenge [71]. The inherent trade-off between sample complexity and generalisation demands ongoing innovation, particularly for real-world tasks with sparse, delayed rewards [24, 54]. Despite these challenges, RL holds potential to introduce the behavioural dynamic that people exhibit when they interact with information during social discourse. When RL is part of a network simulation, rather than imposing static decision rules, agents can enact on reward schemes calibrated to metrics such as conversion probability, uncertainty, and even fact-checking probabilities [104]. RL-augmented infodemic simulations demands attention to model validity, scalability, and interpretability. It is therefore vital that even though a framework for RL in Computational Infodemiology is provided, the architecture and the policies require critical reflection from different perspectives to truly map belief systems. High-dimensional state spaces, such as scale-free networks that represent social media interactions, necessitate function approximation architectures that balance expressivity with transparency [125]. Integrating causal inference techniques into the reward design can help disentangle correlation from causation, permitting counterfactual analyses of intervention strategies [70]. It is for this reason that ABMs and RL can hold a more realistic simulation than ABMs alone—but it must always be approached with caution as this can only map the general belief systems within a population online, and not truly account for what will happen, but rather, what is possible, given human variation.

Another important factor to consider when using RL agents within ABMs is the dependency on creating the correct environment the agents can interact with and what the object of the interactions are (the actions). In an ideal situation, the agents will generate content, and the content would be laden with either truth or misinformation, and such a network can ideally be used to illustrate what would happen if new information is introduced within a society, however, this would at present require a large amount of computational power, methodological validation, and conceptualisation to get right. Therefore, instead of focussing on the actual discourse that may arise within social media echo chambers or online engagement, and trying to model those interactions, it might be more useful to map out belief systems within such a network as a proxy for those that adopt information easily and perpetuate them (such as misinformation) and those agents who are more sceptical. What would also be important to factor into such a model is the change in belief systems, and in the next chapter, this type of belief systems model will be explored.

Chapter 3
A Framework for Modelling Belief Systems

3.1 The Topology of a Belief Map

To conceptualise an infodemic environment for computational infodemic intelligence will require a few informed decisions to implement. First, the network topology, by which it was already justified to be a scale free network to represent social media. Next, the nodes, and what they represent need to be defined. Then, the rules of engagement, and the RL strategies need to be explored. Another important factor is the visualisations of the interactions, the actions, states, rewards, and overall goals of the agents. Finally, even though agents behave unpredictably, some rules within the system are important. These rules will be conditions put into the network to simulate interactions between agents, and their belief states. Let me start with defining what the scale-free network will represent in the context of mapping the belief system within an infodemic.

For belief system mapping, I will focus on one echo chamber at a time, as opposed to the broader social media platform at a specific point in time because it would simply be too complex to map belief systems given all the discourse and influence it represents. An echo chamber will be defined as a semi-closed environment by which a specific series of topics are possible reinforced or viewed based on either a broader topic being discussed (think of it as replies to a statement), or a discourse on a specific topic within a dedicated space (such as the replies to a news feed, comments on a website, comments within a social media page etc.). To map an echo chamber, a scale-free network will be used to mimic the interactions of real-world social media platforms. In social media platforms, a small number of highly influential nodes (influencers) dominate interactions, because they might have more followers, while most of the nodes have limited connections, because they have limited reach. In this type of network nodes will be agents, representing information accompanied by a belief system. Most nodes (agents) have few connections, while a few nodes (hubs or influencers) have many connections. A scale-free network never fully resolves all its nodes, especially when many interactions are present. This means that some

H. Combrink, *Computational Infodemiology*,
SpringerBriefs in Computer Science, https://doi.org/10.1007/978-3-032-10130-3_3

parts, or connections within the network, will not be exposed to the information and its potential reach or spread. If the network was another topology, like a random network, information spreads would be in a relatively uniform manner, which is not the case in a real-world scenario. An important consideration is the belief system of the agents within the network. If RL is to be applied to simulate an agent given an infodemic, the belief systems need to be defined. It is best to understand what the agents need to learn, what actions they need to take, and what their rules are, before it can be done. Therefore, each agent and their belief need to be defined in this multi agent problem. To contextualise this, each node will then represent a belief state, of which, I will start with the simplest of agents, the neutral agents.

3.2 The Agents Within a Belief Map

Neutrals represent individuals who have not yet formed a strong opinion about the validity of a given piece of information. Neutral agents will represent the most susceptible group in the information ecosystem, their belief state is highly malleable, influenced by the agents they interact with. Neutrals serve as swing nodes, capable of transitioning depending on the type of influence they encounter. A neutral is not a person that has not decided what their stance on a topic are, but rather, a person that has never been exposed to it. The primary factor determining whether a neutral belief state changes is exposure frequency. If a neutral agent interacts predominantly with misinformation, they are more likely to adopt misinformation, especially if the exposure occurs through a high-degree node, like an echo chamber reinforcing ideas. Conversely, if they are frequently exposed to interventions, or information, they will not yet know if it is true or not but may likely take on this initial "fact-checking" belief system which may transition toward scepticism of other information. This makes neutral agents key targets to grow a belief network when new information is introduced into the system, as they represent the population most likely to shift the overall network balance. In the early stages of information propagation, neutral agents act as amplifiers, increasing the pool of potential sceptics or believers. A sceptic belief system is fact-checking in nature and slower to adopt new information than a believer. A believer on the other hand assumes a more viral spread of information as the information may be more believable. In the context of the experiments, I will map these belief systems as they underpin information spread as opposed to what information spreads.

In general, believers are the primary drivers of misinformation within the network, actively spreading false or misleading information to their connections. Believers can also spread facts-checking behaviour, but sceptics are generally slower in adopting or changing their point of view. If a believer spreads misinformation, we will call them misfluencers. It is possible to also call an influencer (a node that has a large reach that is spreading misinformation) also a misfluencer. As misinformation spreads, the conversion rate of neutrals into misfluencers often outpaces their conversion into sceptics, leading to misinformation dominance unless countered by strategic interventions. In contrast, sceptics must target neutral agents before they transition into misfluencers, as it is significantly harder to convert a misfluencers than to prevent misinformation adoption in the first place. However, in a scale-free network, certain high-degree nodes significantly amplify specific narratives, and if a misfluencers occupies such a node, misinformation will spread fast throughout the network.

The structure of the scale-free network allows misinformation to propagate quickly if it is at a high-degree node. These agents reinforce misinformation narratives by engaging with other misfluencers, representing the amplification effect of sharing false content, and attempting to recruit neutral agents into their belief system. This loop will continue until the network is occupied. The spread of misinformation follows a probabilistic mechanism where misfluencers attempt to convert neutral agents during interactions. The probability of a successful conversion depends on several factors, including the strength of the misinformation message, the susceptibility of the neutral agent, and the level of exposure within the network. Since scale-free networks feature highly connected nodes, misinformation can spread disproportionately through a few influential believers. In a real-world context, this can be seen when a celebrity or person with a large following also spreads misinformation.

Misfluencers tend to exhibit confirmation bias, reinforcing their commitment to misinformation when faced with contradicting evidence. This is particularly evident when misfluencers interact with sceptics, as they often reject fact-checking efforts outright, viewing them as biased or untrustworthy. Additionally, misfluencers want to use high-engagement strategies to sustain misinformation virality, such as emotional appeals, sensationalism, and repetition, which have been shown to increase the likelihood of misinformation retention and dissemination. This means that being a sceptic requires a specific function, to counter misinformation and neutralise misfluencers. However, in a real-world context, not all misfluencers or sceptics can be "converted" to each other, and thus a certain probability needs to be introduced.

Sceptics serve as the counterforce to misfluencers, actively working to neutralise misinformation by engaging in fact-checking, debunking, and persuading misfluencers or neutral agents to transition toward scepticism. Unlike misfluencers, sceptics propagate information with sources and reduces the level of uncertainty in their information spreading agenda. Sceptics seek to correct misinformation by introducing verified evidence and promoting critical thinking within the network. Their effectiveness depends on both network position and interaction strategy. One of the primary challenges sceptics faces is breaking through misinformation clusters. Believers often form tightly connected groups where misinformation is continually reinforced. In a network, if a sceptic is at the edge of a misfluencers cluster, then a single sceptic will not have the ability to maintain a converted misfluencers, if the misfluencers is connected in a misfluencers cluster. Sceptics must strategically position themselves in the network to maximise their influence and counteract false narratives. Highly connected sceptics have a greater capacity to spread fact-checking information effectively, reaching multiple believers and neutral agents. These high-degree sceptics function similarly to influencers, amplifying corrective information while countering the widespread effects of misinformation diffusion.

The success of scepticism interventions depends on the probability of belief conversion. When sceptics interact with misfluencers, there is a certain probability that they will convince them to transition to a neutral or sceptical state. This depends on the fact-checking probability of the population, the misinformation within the population and most importantly, the uncertainty of the messaging. In a real-world scenario, this probability is influenced by factors such as argument strength, trustworthiness, repetition, credibility, and scientific merit. Sceptics must also contend with backfire effects, where attempting to debunk misinformation can sometimes reinforce a believer's original stance, in other words, when uncertainty is reintroduced into the system. Influencers are therefore the most powerful agents in the network, possessing disproportionate reach and influence over information dissemination. If an influencer is a misfluencer, the messaging amplifies misinformation and the same holds true for sceptics. When influencers are misfluencers, they significantly enhance misinformation virality by broadcasting false information to a wide audience. The presence of high-degree believer influencers leads to rapid misinformation cascades, making counter-interventions difficult because of the scale of the misinformation. Conversely, if influencers are sceptics, they become the most effective debunkers in the network.

3.3 Mapping the Different Belief States

In the experimental framework, each agent in the simulated environment assumes a distinct belief state that governs its orientation toward misinformation. The model defines four principal belief states to capture the dynamics of information diffusion. Agents classified as believers actively disseminate misinformation, contributing to its perceived legitimacy and reinforcing its presence within the network. Sceptics, in contrast, engage in efforts to refute misinformation, aiming to diminish its influence and challenge the agents responsible for its propagation. Neutrals remain passive, refraining from active participation in either the spread or counteraction of misinformation, but retain the potential to transition into either the believer or sceptic states depending on their interactions. A fourth category, influencers, is characterised by elevated connectivity within the network and possesses a heightened capacity to shape belief dynamics. These agents can operate as either misfluencers, amplifying misinformation through widespread reach, or as advocates for factual content, thereby promoting verification and resistance to falsehoods. Belief state transitions among agents are driven by interactions with neighbouring nodes in the network. These encounters lead to probabilistic updates in belief states, influenced by the nature and frequency of exposure to various types of information. The emergent dynamics reflect how social connectivity, information salience, and individual belief predispositions interact to shape collective opinion formation in the context of misinformation. The first being the number of agents within the network.

The number of agents represents the population size in the belief ecosystem and functions analogously to the state space cardinality in RL. In larger networks, the diversity of agent interactions increases, which enriches the agent's experience space and allows for more nuanced emergent dynamics [104]. This parameter affects the complexity of the environment and the potential spread of inforamtion. As the network grows, the RL agent's policy must account for more heterogeneous interactions, mimicking real-world scenarios where misinformation can cascade rapidly through dense or scale-free networks [90]. Therefore, the agent's learning and belief adaptation become contextually dependent on population scale. However, the sample size alone is not enough, and a misinformation spread probability is also needed within the network. Misinformation spread probability reflects the transition probability in RL, particularly governing the likelihood of moving from a less informed to a misinformed state. It models the environment's reward structure by indirectly reinforcing states where misinformation is more likely to be received and adopted. In belief systems, misinformation acts like a pseudo-reward, especially when it confirms existing biases or offers cognitive shortcuts [93]. When the spread probability is high, the environment becomes more hostile to sceptics, and agents require stronger incentives or learning signals to adopt fact-based beliefs. This parameter operationalises the misinformation reward landscape within the RL simulation.

The opposing variable is the fact-checking probability which represents the inverse mechanism to misinformation reinforcement and represents a form of negative reward shaping or resistance to incorrect transitions. It directly affects the belief updating function by providing disconfirming evidence that reduces the perceived value of misinformation-related actions. In RL terms, this is equivalent to reducing the Q-values of misinformative actions, thus encouraging policy shifts away from belief reinforcement based on falsehoods. Fact-checking has been shown to recalibrate belief confidence and can stabilise agent behaviour over time by introducing accurate environmental feedback [69]. This parameter is critical in modelling rational updating within noisy or manipulative environments. Conversion is also not linear and does not depend purely on exposure frequency. Yes, to convert a neutral to either a sceptic or believer is dependent on the information they are initially exposed to, however, once they have a stance there needs to be an indicator that accounts for the possible change that may occur.

The uncertainty probability captures the volatility or decay of learned policies. In RL, this is conceptually like policy drift or the introduction of stochasticity in an otherwise stable belief trajectory. Sceptics, once resistant to misinformation, may revert to neutral if their reinforcement signals weaken or if disinformation exposure persists. This reflects non-stationary dynamics, where the optimal policy changes due to environmental pressure or lack of sustained reinforcement [41]. This parameter ensures the model accounts for belief erosion, cognitive fatigue, or social pressure that may reduce an agent's resistance to false narratives. This parameter is also the most influential in the dynamic. No uncertainty would mean that every agent knows the ground truth, and 100% uncertainty means that no sceptic will ever be able to use any evidence.

The epsilon parameter is derived from the ε-greedy exploration strategy in RL, determining the probability that an agent will explore rather than exploit its current knowledge. In the belief map context, this models the behavioural strategy of believers, particularly their capacity to adapt and try new misinformation tactics. High epsilon values increase exploration, simulating chaotic or opportunistic misinformation dissemination. From a belief formation perspective, this allows agents to sample alternative belief-reinforcement strategies, thus affecting both their local influence and the global dynamics of misinformation spread [89]. The use of epsilon captures the balance between stability (exploitation) and adaptability (exploration) in belief-based behaviours. Each simulation step corresponds to a time-indexed episode in RL, during which agents interact with their environment and update their value functions based on rewards or penalties. Over multiple steps, the belief state progression can be observed, and policy learning can be measured in terms of convergence or divergence. This temporal dimension is essential in RL, as learning is inherently a function of repeated interaction [104]. In belief systems, time is a critical variable, as misinformation effects are often cumulative and non-linear. The number of simulation steps allows for the modelling of both short-term fluctuations and long-term stability or shifts in belief distributions across the network. Each parameter reflects a critical aspect of RL, population complexity (agents), reward transitions (spread probability), adversarial correction (fact-checking), policy instability (uncertainty), exploration dynamics (epsilon), and temporal learning (simulation steps). Together, they form a computational ontology of belief progression that mirrors real-world social, cognitive, and informational influences.

To develop a simulation that integrates smart agents capable of learning throughout a network, it is essential to define their interactions, states, actions, rewards, transitions, and policies. The foundation of this process lies in defining the state of each agent within the misinformation environment, as it dictates their decision-making and adaptive behaviours. In this experiment, an agent's state is characterised by five key factors. Firstly, belief state, categorises agents into one of three groups namely: believer, sceptic, and neutral. Then, local network exposure, which quantifies the proportion of an agent's neighbours that hold different belief states. Thirdly, past interactions have an influence on the state as capturing historical engagement, such as the agent's success rate in spreading misinformation or fact-checking changes the "direction" the agent wants to influence in. The fourth is the influence score, reflecting the agent's centrality within the network, determined by the number of its connections; and finally, misinformation saturation, representing the level of misinformation present in the agent's immediate neighbourhood. These factors collectively define the agent's contextual state, shaping its interactions within the network and influencing its capacity to learn and adapt over time (Fig. 3.1).

In an ideal simulation, node sizes reflect influence, with larger nodes representing influencers who have a greater impact on network-wide misinformation propagation.

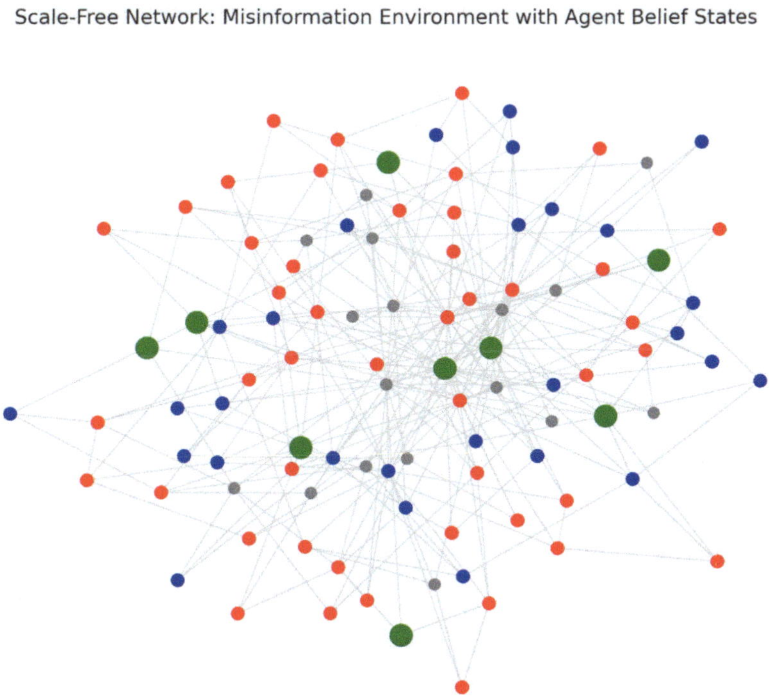

Fig. 3.1 Red: Believers (actively spreading misinformation), Blue: Sceptics (counteracting misinformation), Gray: Neutrals (susceptible to belief shifts), Green: Influencers (high-degree nodes amplifying or mitigating misinformation)

Chapter 4
Modelling Belief Systems

4.1 Introduction

This chapter moves from theory to simulation by implementing the conceptual foundations outlined in the previous chapters through a set of ten computational simulations, at the core of Computational Infodemiology. These simulations are designed to demonstrate how different belief dynamics emerge, evolve, and interact within digitally networked environments. Each simulation functions as a controlled experimental model that captures a unique configuration of social, cognitive, and informational variables, allowing us to examine how belief systems are shaped under varying levels of complexity, uncertainty, and intervention pressure. These simulations were chosen on the basis that, collectively, they represent a broad and valid spectrum of possible belief system trajectories, ranging from environments of moderate information flow to highly adversarial, volatile, or defensive systems. To this end, the number of agents within the network (as a proxy for echo chamber scale and connectivity), the probability of misinformation spread (representing the infectiousness of false narratives), the probability of fact-checking success (modelling resistance to misinformation), a belief instability variable (allowing for reversions or uncertainty in state transitions), the epsilon parameter (governing the degree of exploration vs. exploitation in belief behaviour), and the number of simulation steps (capturing temporal evolution and iterative exposure) were used.

The first imitation, a baseline simulation, offered a reference point with moderate levels of misinformation and limited fact-checking capacity. From there, subsequent simulations incrementally increase the complexity or tension within the system with high misinformation pressure, strong fact-checking networks, high sceptic drop-off probabilities, exploratory believers, passive large-scale networks, and adversarial environments marked by extreme instability and minimal intervention. Of particular interest is the "Mass Chaos" scenario, where an expansive network, high misinformation rates, and weak resistance mechanisms converge to test the system's resilience under collapse conditions. Conversely, simulations like "Community Defence" and

H. Combrink, *Computational Infodemiology*,
SpringerBriefs in Computer Science, https://doi.org/10.1007/978-3-032-10130-3_4

"Tug-of-War" explore cooperative resilience and competitive epistemic equilibrium, respectively. Together, these ten simulations form a belief-system mapping suite that transforms abstract theoretical constructs into tangible, testable models. Each simulation is structured to enable comparative analysis of agent behaviour, belief stability, network influence, and the effect of probabilistic decision-making strategies on emergent social outcomes. The upper confidence bound reinforcement learning framework used across all scenarios ensures that agent behaviours retain the capacity for both habitual exploitation and adaptive exploration.

4.2 Baseline and Structural Simulations

4.2.1 Simulation 1—Baseline

This simulation sets the foundation for comparison, representing a moderate network of 100 agents where misinformation and fact-checking coexist in equilibrium. With a misinformation spread probability of 30% and a fact-checking probability of 10%, the environment reflects a balanced but imperfect system, much like a typical digital community that has access to both unverified content and occasional corrections. As discussed before, the agents that are in a belief state convert faster than a sceptic state. What is important is not the emphasis on the percentage by number (30% for misinformation spread and 10% for fact-checking), but rather that fact-checking is relatively low as compared to misinformation spread. Furthermore, Epsilon is also low (10%), indicating that believers mostly follow familiar paths unless incentivised otherwise. For this simulation, there were 100 agents in the network, and the simulation occurred over 200 steps. Figure 4.1 outlines the final network at the end of the simulation and the model parameters.

Based on the final network, the network is mostly believers, and the believers occupy all but one influencer state. When examining the shift over the duration of the experiment, the believer cumulative reward far outweighs the sceptic reward function (Fig. 4.2).

The population trajectories of believers (red) and sceptics (blue) across the 200 simulation steps. Initially, both groups are relatively balanced, with slight dominance

	Parameter	Value
0	Number of Agents	100
1	Misinformation Spread Probability	30%
2	Fact-Checking Probability	10%
3	Uncertainty	5%
4	Epsilon (E-Greedy)	10%
5	Simulation Steps	200

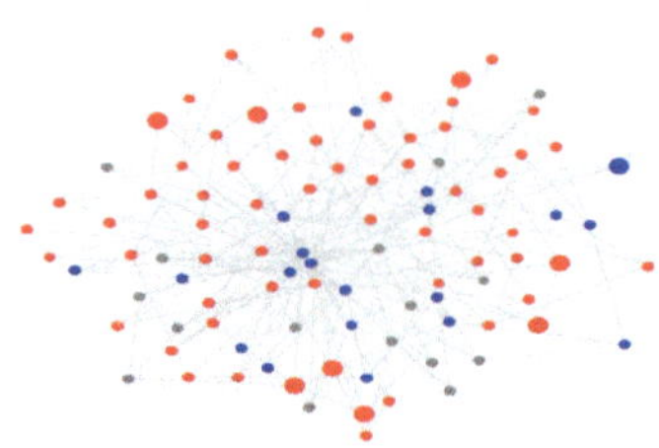

Fig. 4.1 Baseline experiment parameters and network at the end of the simulation

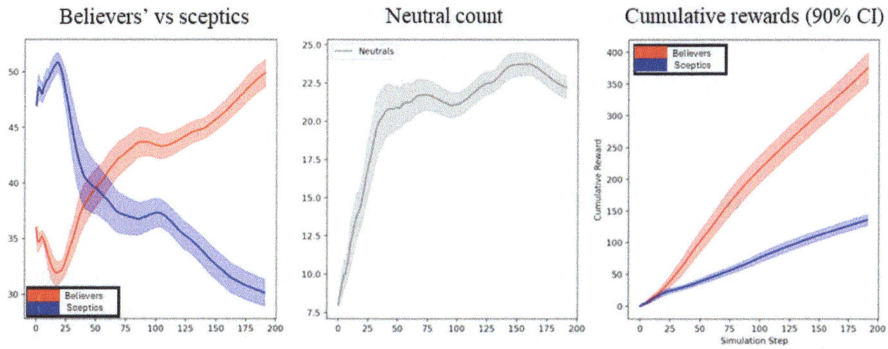

Fig. 4.2 Believer and sceptic simulation trends for baseline experiment

by sceptics. However, a rapid inversion occurs around step 25, where the number of believers begins to rapidly increase, and sceptics decrease. From that point onward, the believer count steadily increases, while sceptic numbers decline sharply. By step 200, believers dominate the network, and sceptics are at their lowest observed level. This trajectory illustrates the long-term effect of asymmetric influence dynamics. These asymmetrical dynamics illustrate that misinformation, with a higher probability of spread and greater cumulative reward, gradually crowds out corrective scepticism in the absence of effective resistance mechanisms. This means that if a network is potentially left to its own devices, that misinformation will spread. This illustration justifies that interventions are needed when misinformation is within a network as the "problem" will not solve itself.

Neutrals increase rapidly in the early phase (steps 0–50), plateauing around step 75 and stabilising thereafter. This suggests an initial surge in disengagement or indecision as agents are exposed to conflicting signals, followed by consolidation. Notably, this plateau aligns with the divergence in the believer and sceptic trends, indicating that as belief polarisation intensifies, the neutral population stabilises. One reason for this could possibly indicate that transitions out of neutrality become increasingly asymmetric (i.e., more likely to convert to believers due to dominant misinformation exposure).

A key insight from the cumulative rewards is that the gap with sceptics widens steadily as the simulation progresses. This is a direct reflection of the model's reward function, where agents gain reinforcement from successfully influencing others. Given the higher misinformation spread probability and the passivity of the system (low ε), believers are more successful in propagating their belief state and are thus more frequently rewarded. Sceptics, on the other hand, receive fewer opportunities to neutralise misinformation or convert others, leading to relatively flat reward growth.

The baseline simulation illustrates how even in a moderately balanced environment, a small advantage in misinformation propagation and limited exploratory behaviour can tip the belief system in favour of misinformed agents over time. The declining sceptic population, stabilisation of neutrals, and growing dominance of believers suggest a fragile equilibrium that ultimately favours misinformation, unless countered by stronger fact-checking forces or more adaptive sceptical strategies. This foundational model sets the stage for deeper comparisons with subsequent simulations that introduce more extreme or resilient belief dynamics.

4.2.2 Simulation 2—Large, Passive Network

In this case, a larger population of 200 agents exists in a relatively low-resistance environment. Fact-checking is rare (5%), while misinformation spread is moderately strong. This mirrors under-moderated or overwhelmed networks (like growing forums or underfunded social platforms), where there's scale without the matching infrastructure to mitigate content distortions (Fig. 4.3).

Simulation 2 offers a clear demonstration of how scale without resistance promotes belief monopolisation. Despite the relatively moderate probability of misinformation spread, the lack of effective countermeasures in a large network result in the near-total collapse of belief heterogeneity. The simulation reflects real-world digital environments in which exponential growth outpaces moderation capabilities, leading to overwhelming misinformation dominance (Fig. 4.4).

	Parameter	Value
0	Number of Agents	100
1	Misinformation Spread Probability	40%
2	Fact-Checking Probability	5%
3	*Uncertainty*	10%
4	Epsilon (E-Greedy)	10%
5	Simulation Steps	300

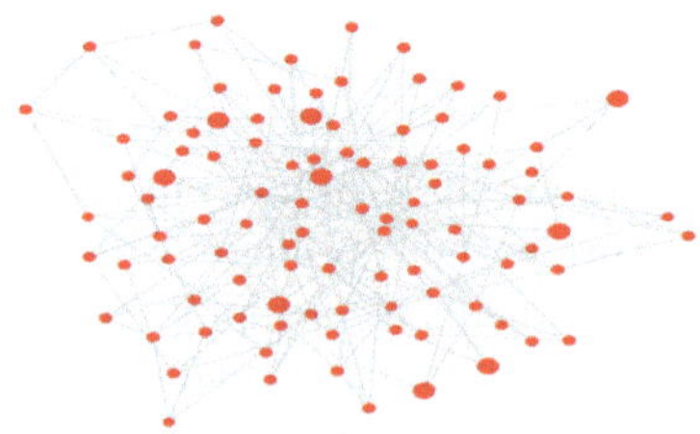

Fig. 4.3 Large, passive network experiment parameters and network at the end of the simulation

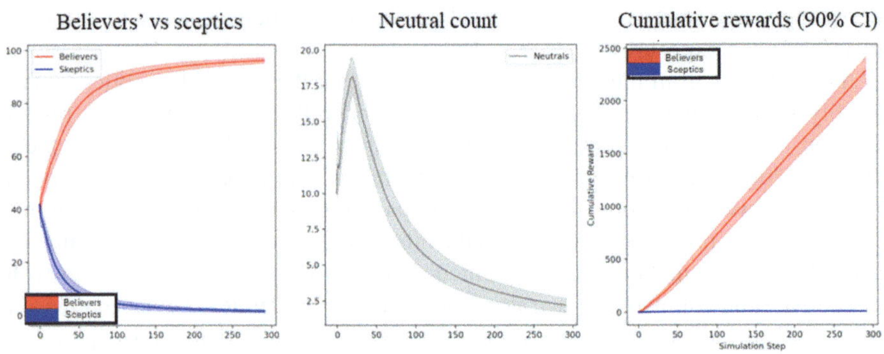

Fig. 4.4 Believer and sceptic simulation trends for large, passive network experiment

This simulation illustrates a rapid and overwhelming dominance of believers. From the initial steps, the proportion of believers sharply increases, surpassing sceptics within the first 25 steps. This early inflection point leads to a compounding advantage, where believers saturate the network. By step 100, sceptics are nearly eradicated, and believers constitute most of the population. The belief dominance curve flattens as it approaches the network's upper bound, indicating saturation. This trend demonstrates how, in the absence of sufficient resistance (i.e., low fact-checking probability), misinformation becomes self-reinforcing, spreading unchecked and converting susceptible agents at scale. The near-zero sceptic population by the end of the simulation signals systemic failure in counter-narrative penetration.

The neutral population rises slightly at the beginning, likely due to initial exposure to conflicting information, but rapidly declines as the system becomes overwhelmed by a unidirectional belief flow favouring misinformation. By step 150, the number of neutrals falls below 5 and continues to decline toward near extinction. This behaviour reflects a progressive collapse of belief diversity, as agents are increasingly forced into alignment with the dominant misinformation narrative due to the absence of credible alternatives or corrective feedback. The loss of neutral ground signals a fully polarised ecosystem where informational nuance is eliminated. The cumulative reward panel captures a striking divergence in agent performance. Believers accumulate rewards at an exponential rate, with a steep, consistent trajectory over the 300-step simulation. In stark contrast, the reward curve for sceptics is virtually flat, indicating negligible influence or impact within the environment. The vast difference between the reward accumulation paths, even within the 90% confidence intervals, reveals the asymmetric reinforcement dynamics at play where misinformation behaviours are repeatedly and reliably rewarded, while fact-checking agents lack both the opportunities and mechanisms to gain traction or credibility. This reward imbalance further entrenches misinformation dominance and underscores the systemic vulnerability of under-moderated networks. Importantly, the model reveals that size alone does not guarantee diversity or resilience. Without active and systemic interventions, belief systems in large passive networks are highly susceptible to being overtaken by the most aggressively reinforced narratives.

4.2.3 Simulation 3—Deterministic Behaviour

Believers in this setup don't explore at all (epsilon = 0.0), representing strong confirmation bias or filter bubble dynamics. The environment itself is relatively neutral, but the behavioural rigidity models agents who only reinforce existing beliefs and connections. This could simulate closed communities or algorithmic echo chambers where opposing views are minimised (Fig. 4.5).

A clear and consistent decline in the population share of believers over time, while sceptics illustrated a mirrored, steady increase (Fig. 4.6).

Initially, the proportions of the two groups are comparable, but a strong divergence emerges within the first 50 simulation steps. By the end of the 200-step period,

	Parameter	Value
0	Number of Agents	100
1	Misinformation Spread Probability	30%
2	Fact-Checking Probability	30%
3	*Uncertainty*	5%
4	Epsilon (E-Greedy)	0%
5	Simulation Steps	200

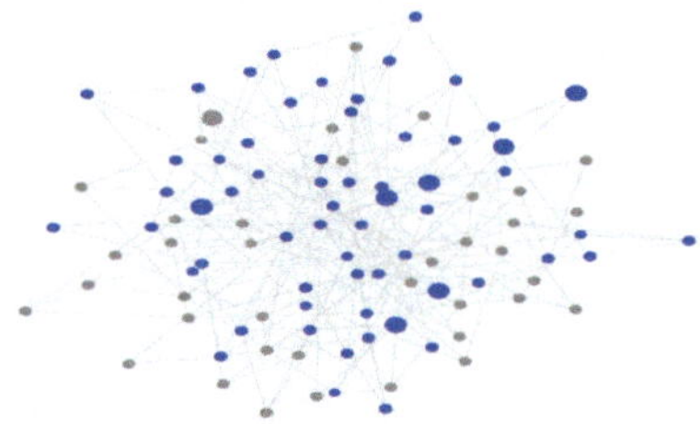

Fig. 4.5 Deterministic behaviour experiment parameters and network at the end of the simulation

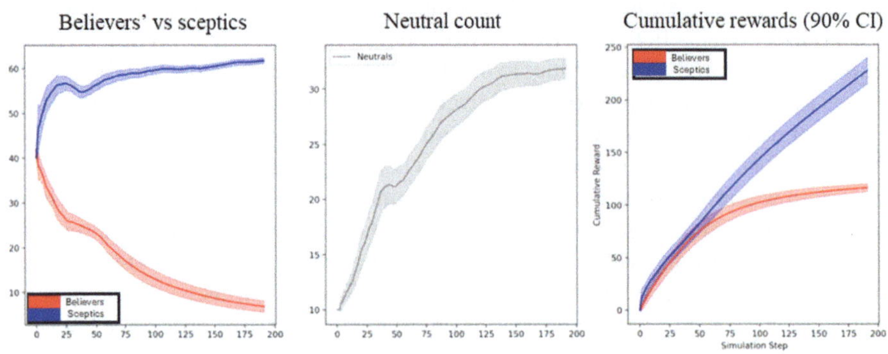

Fig. 4.6 Believer and sceptic simulation trends for deterministic behaviour

sceptics constitute the dominant faction, whereas believers have nearly vanished. The tight confidence intervals suggest minimal variance across simulation runs, reinforcing the interpretation of deterministic system dynamics. In this simulation, outcomes are not shaped by stochasticity but by structural features of the environment or decision policy. The neutrality gradually increases in number before plateauing. This implies a slow but stable absorption of agents into a neutral stance, possibly due to indecision, reduced payoff sensitivity, or structural transitions that delay commitment to network extremes. Unlike the previous simulations, this illustrated how networks rapidly swing when exploration is removed from believers. The sceptics consistently outperform believers in cumulative reward, with the divergence becoming pronounced after approximately 50-time steps. Taken together, these figures present compelling evidence for deterministic behavioural convergence in the belief ecosystem. The dominance of sceptics can be understood as an emergent property of the system's reward structure, which systematically favours sceptical agents when the ability of "new information" in the system is removed. This property suggests that belief systems, even though they dominate in baseline experiment, require creativity to flourish. Fact-checking might take longer but is more rigid and less mailable to creative change. What is important to distinguish in these experiments is that exploration is based on seeking the most mailable, and not "what information is the best to spread the misinformation". Over time, and in the absence of significant noise or external shocks, agents rationally gravitate toward this more advantageous state.

4.3 Misinformation Conflict Simulations

4.3.1 Simulation 4—Strong Fact-Checking Network

This scenario keeps misinformation at a moderate level but increases fact-checking probability significantly to 0.4, creating an environment resembling proactive communities or platforms where content moderation and fact-checking are emphasised. The agent behaviour remains otherwise unchanged, meaning the key variable being modelled is the efficacy of rapid and frequent intervention against misinformation (Fig. 4.7).

The model offers a striking contrast to the high-misinformation environment previously discussed, revealing how structural epistemic interventions can significantly alter the trajectory of belief evolution (Fig. 4.8).

	Parameter	Value
0	Number of Agents	100
1	Misinformation Spread Probability	30%
2	Fact-Checking Probability	40%
3	Sceptic Conversion Probability	5%
4	Epsilon (E-Greedy)	10%
5	Simulation Steps	200

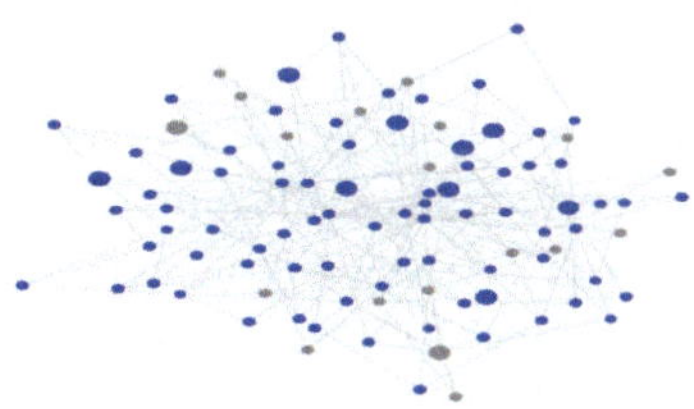

Fig. 4.7 Strong fact-checking network experiment parameters and network at the end of the simulation

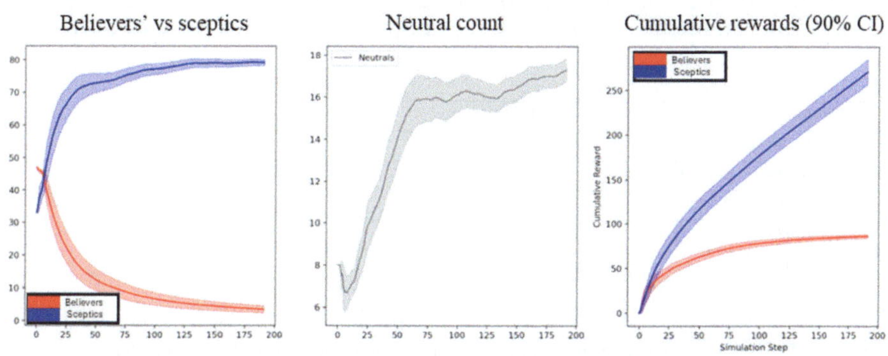

Fig. 4.8 Believer and sceptic simulation trends for strong fact-checking network

The simulation of agent states in this environment indicates a dramatic reversal of belief dominance. At the outset, believers and sceptics are roughly equal in population. However, believers experience a rapid and sustained decline, falling from over 60 to below 5 by the end of the simulation. In contrast, sceptics grow steadily, reaching approximately 80% of the population and stabilising near the end. The narrow confidence intervals once again signal a deterministic convergence, with low variance across simulation runs. This pattern is consistent with an environment in which fact-checking mechanisms are both effective and pervasive. Agents who begin with belief-based stances are exposed to corrective feedback and, through reinforcement learning and a reward-driven policy, gradually shift toward scepticism. This is not a stochastic fluctuation but a systematic transformation in epistemic state, underscoring the structural power of information verification. The problem is, if the facts themselves are misinformed, this becomes a major concern. Networks like this rely on good information because fact-checking behaviour is much harder to sway away from false information that was deemed as the source.

4.3.2 Simulation 5—Mass Chaos

With 300 agents, high misinformation spread (0.7), and weak fact-checking (0.05), this simulation depicts a **chaotic information ecosystem**—such as early stages of a crisis, geopolitical disinformation campaigns, or unmoderated platforms at scale. It models how fast and deeply misinformation can saturate a network when constraints and corrections are insufficient (Fig. 4.9).

With 300 simulation steps, the scenario likely simulates environments such as early-stage crises, geopolitical disinformation campaigns, or unmoderated social media platforms during breaking news events. The simulation thus not only reflects a volatile environment but also asymmetric decision-making strategies under information pressure. This experiment was repeated with similar, but unpredictable outcomes (Fig. 4.10).

This plot displays oscillatory, co-fluctuating behaviour between believers and sceptics. Both groups hover between 36 and 48 agents throughout the 300-step period,

	Parameter	Value
0	Number of Agents	100
1	Misinformation Spread Probability	70%
2	Fact-Checking Probability	15%
3	*Uncertainty*	5%
4	Epsilon (E-Greedy)	10%
5	Simulation Steps	300

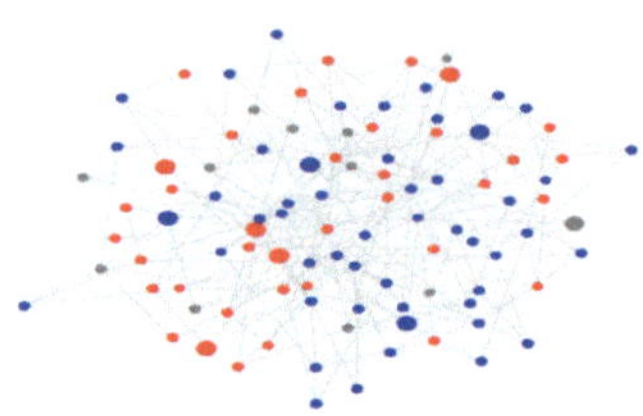

Fig. 4.9 Mass chaos experiment parameters and network at the end of the simulation

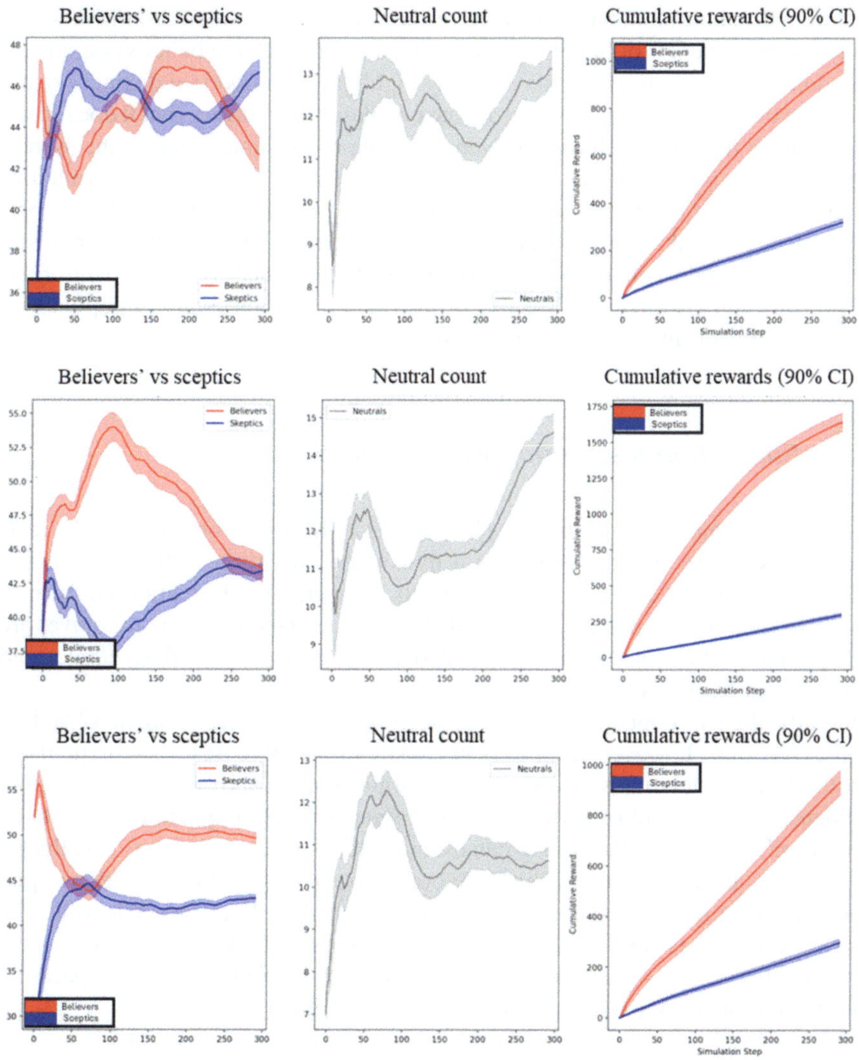

Fig. 4.10 Believer and sceptic simulation trends for mass chaos

with frequent crossovers and relatively wide confidence intervals, unlike the deterministic outcomes of prior simulations. This fluctuation points to a fragile equilibrium where neither group consolidates dominance. The believer population tends to peak slightly above the sceptics, but not consistently or decisively. These dynamics are indicative of a turbulent cognitive environment, where the absence of strong correction mechanisms allows belief and scepticism to compete without resolution. Despite the equilibrium in population counts, believers significantly outperform sceptics in cumulative rewards, nearing 1000 by the final timestep. Sceptics lag far behind, reaching only about 400. This growing reward gap reflects an environment structurally biased in favour of belief reinforcement, even though the belief system itself never dominates numerically. This disparity highlights an important distinction: the reward structure does not necessarily dictate population dominance, but it does shape the incentives that drive behaviour. Even in a state of dynamic balance, the system disproportionately rewards belief-aligned strategies (likely through misinformation-driven engagement or reinforcement loops), ensuring that believers remain competitive in decision-making and strategy evolution. This simulation offers a realistic portrait of epistemic volatility in environments characterised by high misinformation pressure and insufficient corrective infrastructure. In such contexts, populations do not converge cleanly toward truth or falsehood.

4.3.3 Simulation 6—Tug-of-War Scenario

A relatively balanced configuration (150 agents, 40% misinformation, 30% fact-checking, 10% uncertainty) simulates a dynamic equilibrium, where both belief systems are in active contention. This reflects competitive information landscapes like election seasons or policy debates, where accurate and misleading narratives are equally resourced and persuasive, and outcomes depend on network dynamics and reinforcement loops (Fig. 4.11).

Initially, the two populations are near parity, each oscillating between 42 and 44 agents. Over time, however, sceptics begin to slightly outpace believers, particularly after the 100th simulation step, though the margins remain narrow. The overlapping confidence intervals and recurring crossover points suggest a dynamically unstable equilibrium, wherein neither group can establish long-term dominance. This fluctuating behaviour indicates that agents are continuously exposed to competing epistemic pressures, misinformation remains influential but is frequently checked. The close tracking of these two groups is emblematic of systems where epistemic contestation is active and sustained, such as public debates in regulated media spaces or contested narratives during slowly unfolding crises (Fig. 4.12).

	Parameter	Value
0	Number of Agents	100
1	Misinformation Spread Probability	40%
2	Fact-Checking Probability	30%
3	*Uncertainty*	10%
4	Epsilon (E-Greedy)	10%
5	Simulation Steps	250

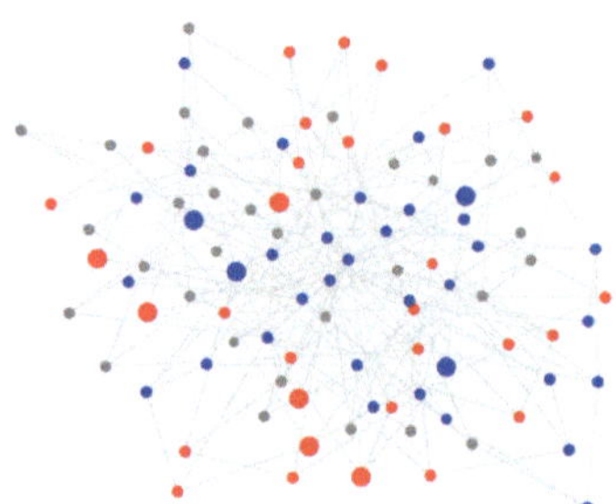

Fig. 4.11 Tug-of-war scenario experiment parameters and network at the end of the simulation

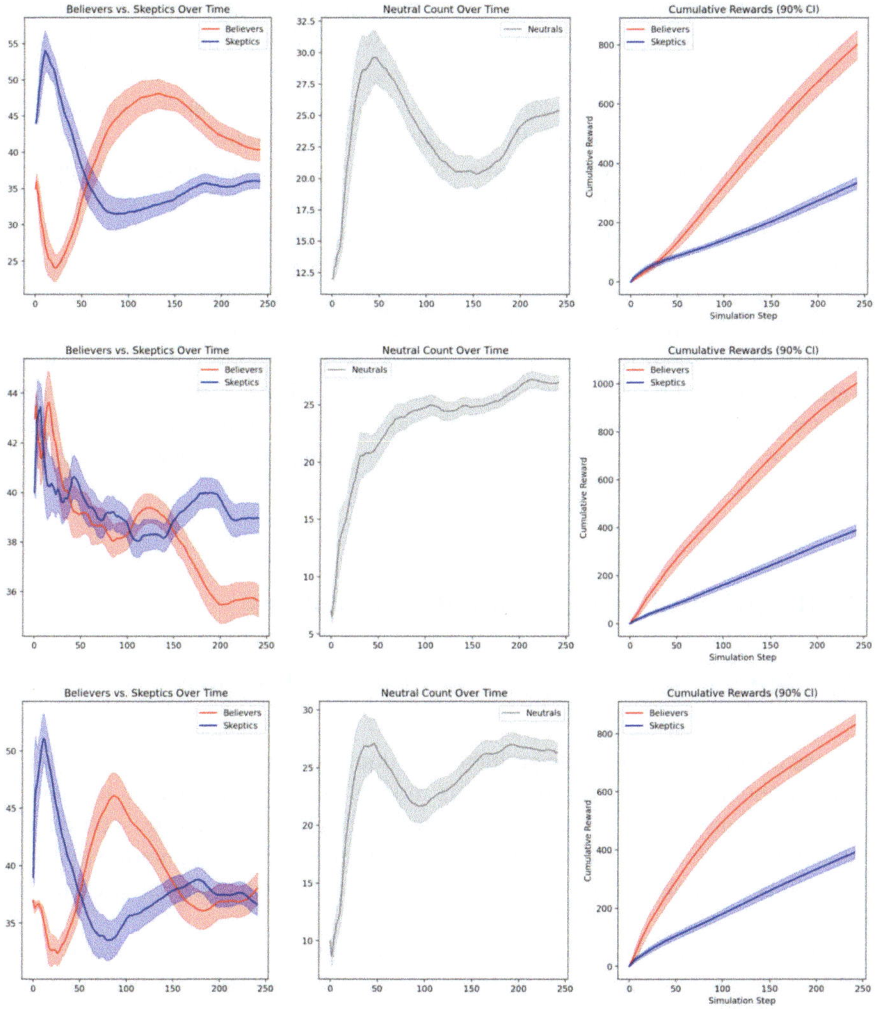

Fig. 4.12 Believer and sceptic simulation trends for tug-of-war scenario

This simulation exemplifies a contested epistemic space, where neither misinformation nor correction fully prevail, but the environment is nevertheless tilted toward rewarding belief propagation. The increasing neutral population reflects a society grappling with competing narratives and opting out of binary positions. The fluctuation between believers and Sceptics suggests that even marginal changes in information flow can tip the balance in such systems.

Across distinct scenarios a consistent pattern emerges where the architecture of rewards and the balance of epistemic inputs are far more determinative than agent predispositions alone. Even in systems where scepticism is numerically stable or growing, belief-based strategies often flourish when they are systematically rewarded, underscoring a critical decoupling between truth and utility. The dynamics of neutral agents highlight the growing cognitive strain or informational fatigue agents face when exposed to equally persuasive yet opposing narratives. These agents represent not indecision, but a reflective pause within chaotic systems. Moreover, oscillatory behaviour between believers and sceptics in turbulent settings suggests that equilibrium is often illusory without structural correction. These system remains vulnerable to cycles of misinformation and fragile consensus. What these findings suggest is that epistemic ecosystems are inherently shaped by the rules of engagement. Whether through algorithmic design, reinforcement policies, or social incentives, the flow of information and the structure of rewards forge the path that belief systems follow. When rewards favour veracity, belief systems self-correct. When rewards favour virality, systems drift toward distortion. These insights open a crucial space for rethinking the architecture of digital platforms, policy interventions, and educational approaches in the age of high-velocity misinformation because there is a sense of self correction.

4.4 Behavioural and Community Simulations

4.4.1 Simulation 7—High Misinformation Pressure

Here, the misinformation spread probability is doubled (0.6) while fact-checking and sceptic dropout rates remain steady. This models an ecosystem where misleading content is aggressively circulated, perhaps through viral sharing, coordinated campaigns, or platform algorithms, overwhelming standard countermeasures. A mid-sized population implies enough complexity for cascading effects without total saturation (Fig. 4.13).

This dynamic simulates that misinformation thrives in this environment where there is a relatively high probability that overwhelms weak corrective influence of

	Parameter	Value
0	Number of Agents	150
1	Misinformation Spread Probability	60%
2	Fact-Checking Probability	10%
3	Uncertainty	5%
4	Epsilon (E-Greedy)	10%
5	Simulation Steps	200

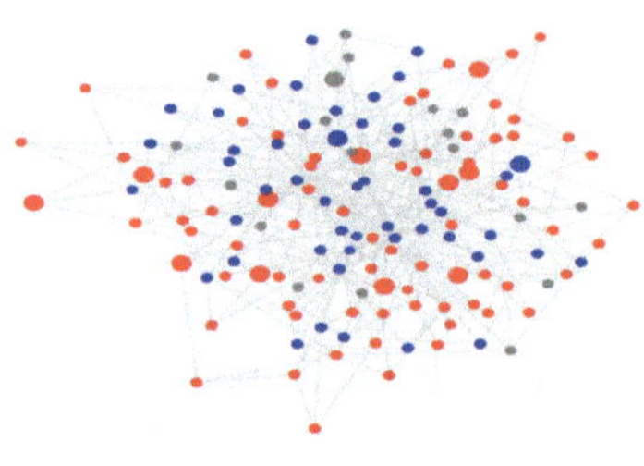

Fig. 4.13 High misinformation pressure experiment parameters and network at the end of the simulation

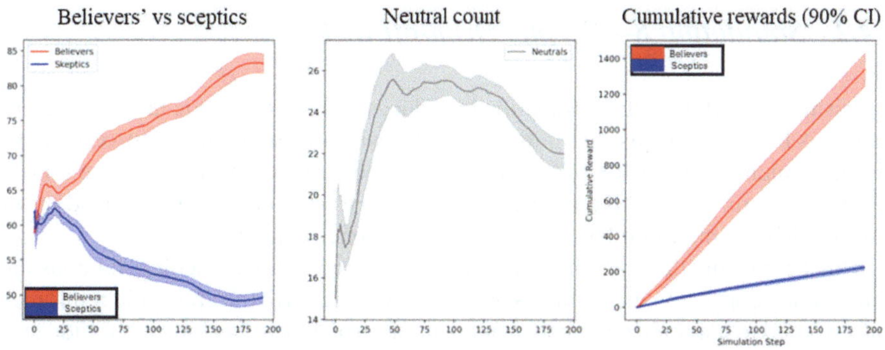

Fig. 4.14 Believer and sceptic simulation trends for high misinformation pressure

fact-checking. Once misinformation begins to spread, its momentum compounds over time, driving agents toward the belief-oriented state (Fig. 4.14).

In this simulation, the number of believers increases sharply, rising from approximately 60 to over 85 by the end of the simulation. In contrast, the sceptics decline steadily, dropping below 50, with minimal fluctuation. The steep divergence begins early in the simulation and continues monotonically, indicating strong reinforcement dynamics. The proportion of neutral agents rises rapidly in the early stages, peaking at around 75 steps, but then declines steadily through to the end. This inverted-U trajectory suggests that while uncertainty initially prompts a sizable number of agents to adopt a neutral or undecided position, the system quickly polarises. The decline in neutrals aligns with the growing dominance of Believers, implying that uncertainty is resolved in favour of belief as misinformation proliferates unchecked. This observation highlights how neutrals can serve as transitional agents, buffering the population during early dynamics but ultimately being absorbed into dominant ideological poles. In this case, the low uncertainty parameter likely accelerates belief fixation, compressing the time window during which neutrality is viable. The reward structure clearly favours believers, whose cumulative rewards increase sharply and linearly over time, ultimately exceeding 1400. Sceptics, by contrast, show a nearly flat growth curve, plateauing below 300. This stark divergence reinforces the strategic advantage of adopting a belief-driven stance in a high misinformation environment.

This simulation under high misinformation pressure demonstrates how asymmetric reinforcement mechanisms, coupled with weak fact-checking and limited uncertainty, can rapidly polarise a population toward belief. The convergence pattern reflects not merely belief diffusion but a structural reward asymmetry that reinforces misinformation pathways. The short-lived presence of neutrals suggests limited deliberation or information processing, further exacerbated by the low uncertainty and exploration rate. These dynamics point to the fragility of epistemic diversity in environments where misinformation is highly transmissible and epistemic correction mechanisms are underpowered. From a systems perspective, the simulation highlights the dangers of informational imbalance, where the architecture of influence structurally favours low-veracity content.

4.4.2 *Simulation 8—Sceptic Drop-off Risk*

With a high uncertainty, this simulation introduces the idea of sceptic fatigue. Sceptic fatigue is based on agents that initially resist misinformation but are susceptible to disengagement or confusion over time. The structure reflects a situation where fact-checking agents operate, but their consistency or trust is compromised, leading to attrition. It's an important lens on long-term resilience (Fig. 4.15).

This behaviour illustrates how high uncertainty accelerates ideological consolidation when correction mechanisms are weak. Sceptics lack the epistemic reinforcement needed to maintain coherence, and without strong fact-checking or confidence-building signals, scepticism deteriorates into either belief adoption or disengagement (Fig. 4.16).

This simulation exemplifies a fragile epistemic system where uncertainty accelerates ideological collapse. Unlike high-misinformation environments that overwhelm through volume, this configuration allows belief systems to prevail not through force, but through the absence of viable alternatives. Sceptics drop off not because they are outargued, but because they are under-supported, both informationally and structurally.

	Parameter	Value
0	Number of Agents	100
1	Misinformation Spread Probability	30%
2	Fact-Checking Probability	10%
3	Uncertainty	20%
4	Epsilon (E-Greedy)	10%
5	Simulation Steps	200

Fig. 4.15 Sceptic drop-off risk experiment parameters and network at the end of the simulation

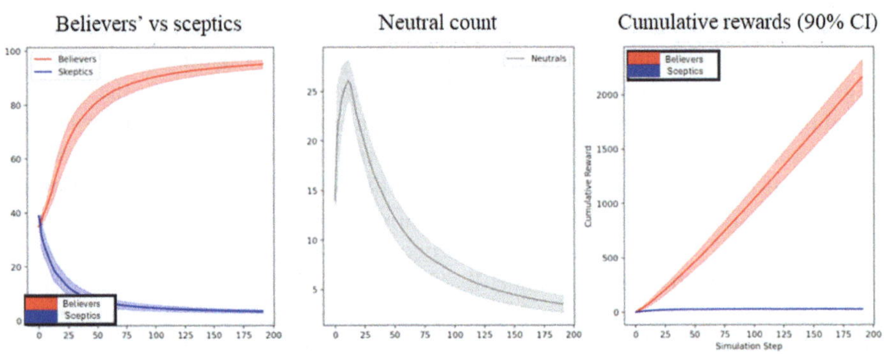

Fig. 4.16 Believer and sceptic simulation trends for sceptic drop-off risk

4.4.3 Simulation 9—Exploratory Believers

What's being represented here is a population where believers are highly exploratory. This could reflect environments of information overload or novelty-seeking behaviour, where users are more likely to click on unknown or unverified content out of curiosity or exposure bias (Fig. 4.17).

This simulation captures the dynamics of an information environment characterised by low misinformation pressure, and moderately strong fact-checking, yet significantly limited capacity for sceptical conversion (Fig. 4.18).

The results reveal a rapid and stable dominance of sceptics, whose population grows from parity to over 90% within the first 40 steps, coinciding with a sharp and irreversible decline in believers. The reward structure reinforces this outcome, with sceptics accruing substantially more cumulative utility over time. This suggests that, in this system, truth-aligned behaviour is both socially reinforced and structurally incentivised. Interestingly, the neutral population grows slowly and stabilises at a low level, indicating that uncertainty remains minimal in such a well-regulated informational ecosystem. This pattern mirrors real-world conditions where proactive verification mechanisms and constrained misinformation pathways lead to epistemic resilience, favouring scepticism without provoking cognitive overload or disengagement. However, the persistently low sceptic conversion rate implies that once

	Parameter	Value
0	Number of Agents	100
1	Misinformation Spread Probability	20%
2	Fact-Checking Probability	30%
3	*Uncertainty*	2%
4	Epsilon (E-Greedy)	10%
5	Simulation Steps	150

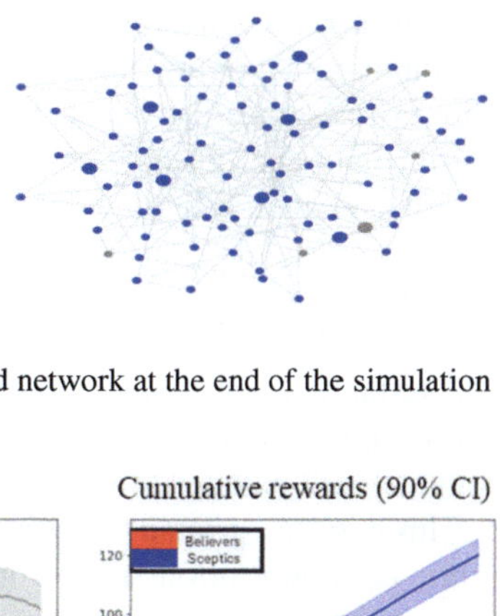

Fig. 4.17 Exploratory believers experiment parameters and network at the end of the simulation

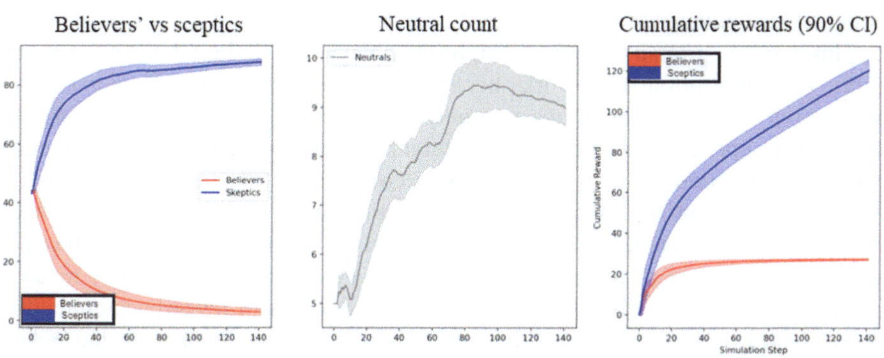

Fig. 4.18 Believer and sceptic simulation trends for exploratory believers

entrenched, belief systems may become resistant to change. This also means that this type of network dynamic requires ongoing structural incentives rather than relying on individual cognitive recalibration.

The results presented across these simulations offer a compelling and multidimensional view of how belief systems evolve under conditions of varying informational integrity, structural incentives, and behavioural uncertainty. Each configuration reveals responses shaped by the nuanced interplay of structural parameters. At the core of this analysis lies the consistent finding that belief system dominance is not inherently tied to informational accuracy or truth value, but rather to the reward structures embedded within the system. When cumulative reward consistently accrues to agents aligned with belief, regardless of whether that belief is grounded in truth or misinformation, the system begins to self-reinforce around that strategy. This aligns with behavioural economics and experimental economic simulations [14]. This is most visibly demonstrated in the high misinformation pressure scenario, where the cumulative rewards of believers vastly outpace those of sceptics, given slight changes in fact-checking and exploration. Even when neutrals initially rise in response to uncertainty, they are quickly absorbed into the more structurally advantageous state. Conversely, in scenarios characterised by strong fact-checking and high sceptic reward (e.g., the strong fact-checking environment), the system converges toward a stable sceptical majority. This shift is not simply due to informational abundance or the superiority of rational argument, but rather because sceptic-aligned strategies offer more use for the system over time, making belief adoption suboptimal in a reward-maximising framework. In these environments, neutrals either remain marginal or are quickly drawn into the sceptical camp, with minimal fluctuation or resistance. However, the most conceptually rich insights emerge from the intermediate and unstable cases, particularly those marked by high uncertainty or reward asymmetry. In the "sceptic drop-off risk" scenario, for instance, we observe a fragile epistemic ecosystem in which the reward advantage of believers and the elevated uncertainty threshold led to a rapid attrition of sceptical agents. Notably, this occurs despite moderate misinformation levels, indicating that uncertainty alone can destabilise epistemic pluralism if the architecture of rewards favours belief. In this configuration, neutrals initially surge, reflecting deliberative hesitation, but ultimately migrate toward belief.

This case foregrounds a critical insight often neglected in digital governance and public communication frameworks because it is not just the volume of misinformation, but the insufficiency of correction coupled with high cognitive ambiguity that leads to epistemic breakdown. A different but equally significant phenomenon is observed in simulations where epistemic competition persists. For example, in chaotic information spaces marked by fluctuating agent states and co-dominance of belief and scepticism, the system never achieves epistemic consensus. Instead, it settles into a state of turbulent equilibrium, where both belief systems retain a foothold, yet believers disproportionately benefit in terms of reward accumulation. In real-world terms, this scenario mirrors online platforms where truth and misinformation coexist, yet the latter gets more clicks, visibility, or engagement due to reward-maximising algorithms.

This introduces an important theoretical tension between epistemic stability and epistemic plasticity, suggesting that too little belief volatility may entrench dogmatism, even in truth-aligned systems. Thus, maintaining epistemic health may require a calibrated degree of belief permeability, such that agents can revise positions without systemic inertia or ideological ossification. Across all simulations, the role of neutral agents emerges as an important diagnostic for system health. In environments of high uncertainty and reward imbalance, neutrals tend to be transient and unstable. Uncertainty also deeply disrupts these systems. However, in more stable environments, their presence diminishes not through suppression but through rational resolution. The implication is that neutrality is a fragile but necessary buffer, capable of absorbing shocks and delaying premature convergence. Its survival is conditional on a system that supports deliberation, tolerates uncertainty, and refrains from over-rewarding polarised strategies. Moreover, the simulations suggest that resilience is not the absence of belief fluctuation, but the ability of a system to contain and guide those fluctuations toward truth-convergent pathways through well-aligned structural rewards. These findings carry important implications for digital platform design, public communication strategies, and misinformation policy. They suggest that algorithmic transparency, value-sensitive reward design, and strategic uncertainty modulation are not merely add-ons, but core components of epistemic governance in the age of information abundance. As belief systems are increasingly shaped not by content alone but by the systemic architecture of attention, feedback, and reward, the design of these systems becomes an ethical imperative. Without such systemic consideration, even moderate levels of misinformation, when paired with weak correction and high uncertainty, can tilt entire populations toward epistemic collapse. Ultimately, the nuanced patterns identified here point to a broader research agenda for Computational Infodemiology, one that does not seek singular truth states but interrogates the structural affordances that make some truths survivable and others extinguishable. Through these simulations, we see not just belief systems in motion, but the contours of belief ecologies, fragile, adaptive, and profoundly shaped by design.

Chapter 5
Social Stress Indicators

5.1 Social Stress Indicators and Infoveillance

The measurement of social stress within online environments has become a critical aspect of digital discourse analysis, particularly in the context of misinformation propagation, ideological polarisation, and heightened emotional contagion in online interactions. To track and understand what is happening online is a major gain, not just in the context of public health and business, but also as an early warning detection system and a very important aspect of digital infoveillance. Social stress indicators (SSIs) provide a computational framework to assess the intensity, prevalence, and patterns of indicators that may be linked to different types of sentiment, stress-related dialogue, conversations of concern, and/or monitoring different topics, contexts, and digital platforms that may lead to some form of social or individual harm. As with any indicator, a certain level of quantitative measurement is required to track the indicator. As online conversations shape public perception, influence decision-making, and contribute to societal tensions, the ability to measure social stress dynamically offers insights into the psychological and sociological dimensions of digital communication. This chapter explores the foundational principles of social stress indicators as a methodological tool for understanding and quantifying social stress in online discourse, the first one being search volume indices as a proxy for information seeking behaviour.

The concept of social stress in digital environments is multifaceted, encompassing elements of emotional distress, cognitive overload, and behavioural responses to online stimuli. One area of social stress however that remains largely unexplored is the potential harm and potential stress that certain messaging may cause. From an infoveillance perspective this gap has been explored with a variety of different ideas. Unlike traditional stress models, which focus on individual psychological responses, SSIs extend this analysis to a broader, population-level scale by integrating computational techniques such as sentiment analysis, engagement metrics, and information-seeking behaviours. The theoretical underpinning of social stress

H. Combrink, *Computational Infodemiology*,
SpringerBriefs in Computer Science, https://doi.org/10.1007/978-3-032-10130-3_5

indicators is grounded in the recognition that online discussions do not exist in isolation but are deeply intertwined with the dynamics of misinformation spread, social amplification, and digital virality. As users engage in online discourse, their interactions, whether through passive exposure, active participation, or resistance, contribute to the emergent patterns of social stress. Measuring these interactions in real time provides valuable insights into how stress levels fluctuate in response to specific events, crisis moments, or high-engagement topics.

Not all topics generate the same level of stress or stress potential. Different online communities also exhibit varying sensitivities to specific issues. For example, discussions around public health crises, political polarisation, or financial uncertainty tend to evoke higher stress responses compared to routine or neutral topics. To this extent, an SSI or several different SSIs can allow researchers to quantify these variations systematically. By capturing these nuances, SSIs serve as a valuable tool for detecting potential points of social tension before they culminate in widespread digital panic, misinformation cascades, or online hostility. In a broad sense, SSIs can be used as a form of infoveillance, a surveillance technique using indicators and social listening to monitor discourse. To "listen" to high frequency social media and other digital media is unrealistic as there simply is not enough time to critically read and respond to all the available content. However, by implementing some of the SSIs, infoveillance is possible to flag potential areas of concern once certain thresholds are reached.

Furthermore, the adaptability of SSIs to different digital ecosystems underscores their relevance for Computational Infodemiology. Social media platforms, news aggregation sites, and online forums each present unique challenges in measuring stress, requiring indicators that can dynamically adjust to different communication structures and network topologies. The degree to which stress proliferates within an online network is influenced by the connectivity of users, the presence of influencers, and the virality of emotionally charged content. In this regard, SSIs not only serve as a diagnostic tool for measuring stress but also as a predictive mechanism for forecasting shifts in digital discourse, enabling researchers and policymakers to intervene in online discussions before stress escalates to harmful levels. The extent to which SSIs can be used to predict future behaviour needs to be explored, but at the very least, social discourse through SSIs can be tracked.

The development and implementation of SSIs require careful consideration of computational methodologies, including natural language processing (NLP), machine learning algorithms, and behavioural analytics. Sentiment analysis, a core component of SSIs, helps identify the emotional polarity of digital conversations, distinguishing between positive, negative, and neutral tones. Additionally, information-seeking behaviour, measured through search trends, content engagement metrics, and keyword associations, provides further contextual grounding for assessing social stress in digital environments. These computational approaches allow for the aggregation of social stress metrics over time, enabling a longitudinal understanding of how stress patterns evolve and interact with misinformation propagation. High levels of social stress are often correlated with sensationalised narratives, as emotionally charged content is more likely to be shared, reacted to, and amplified within online networks. This amplification effect has significant implications for misinformation research, as stress-laden discussions are particularly susceptible to cognitive biases, echo chamber effects, and viral misinformation spread.

Physical community engagement is still needed. SSIs alone are not enough to monitor and evaluate crisis communication, but given the high volume of social discourse, and the sensationalism in social media, SSIs are becoming more important. As online environments continue to shape public opinion, drive policy debates, and influence societal cohesion, the ability to measure stress dynamically is essential for developing resilient and informed digital communities.

5.2 Search Volume Indices

Search volume indices is a normalised indicator derived from a platform where people query content through a recommender system that quantifies the relative frequency with which a particular keyword search or phrase over a specific temporal and geographic scope. The normalisation process ensures that the index is scaled independently of absolute search counts, thereby controlling for fluctuations in overall search engine usage over time. This enables temporal comparisons by mitigating distortions introduced by the expanding user base or increasing internet penetration. While search volume indices do not capture every detail of why someone searches, they are widely accepted in research as a useful proxy or stand-in for understanding information-seeking behaviour (ISB). In essence, when more people search for a topic, it reflects greater public curiosity or concern about that subject. This becomes especially important when we consider how social norms are formed. According to the three-stage model of social norm formation, individuals move through three phases, (1) information acquisition, (2) attitude formation, and (3) behavioural reinforcement. The first stage, information acquisition, is where search volume indices become most relevant. In this phase, people are unsure about what to believe or how to act. They actively seek out information, ask questions, compare viewpoints, and try to understand the new or evolving issue. This surge in information-seeking activity is often visible in a rise in online search behaviour.

By tracking ISB during this critical early stage, researchers can detect when a population is forming new attitudes or beliefs, before those beliefs solidify into long-term norms or behaviours. In other words, search volume indices not only reveal what people are curious about, but when they are most open to influence. This makes ISB a powerful tool for anticipating public opinion trends, tailoring communication strategies, and designing timely interventions in areas such as public health, misinformation control, and social awareness campaigns.

5.3 Natural Language Processing (NLP): Sentiment Analysis

Natural Language Processing (NLP) is a subfield of artificial intelligence (AI) that focuses on enabling computers to understand, interpret, generate, and respond to human language in a meaningful way. By using computational techniques, statistical models, and linguistic rules, NLP facilitates the automated processing of vast amounts of text-based data, making it particularly valuable for analysing large-scale digital communication, including social media discourse. NLP encompasses a range of tasks such as tokenisation, syntactic parsing, named entity recognition (NER), part-of-speech (POS) tagging, and semantic analysis. The application of NLP in online environments allows researchers to extract patterns from unstructured text data, revealing underlying trends in communication, emotional expression, and information flow. Using machine learning and deep learning models, NLP can classify, summarise, and assess text, making it a critical tool for computational social science, infodemic intelligence, and the study of social stress indicators.

Among the various tasks within NLP, sentiment analysis plays a crucial role in understanding the emotional tone and attitude embedded within digital conversations. Sentiment analysis, also known as opinion mining, involves the systematic evaluation of text to determine whether the expressed sentiment is positive, negative, or neutral. This computational approach relies on supervised learning models trained on labelled datasets, lexicon-based methods that match words to predefined sentiment scores, and hybrid techniques that combine rule-based heuristics with machine learning classifiers. Sentiment analysis is widely used in social media research, customer feedback evaluation, political discourse analysis, and public sentiment tracking, as it enables the quantification of emotional responses on a large scale. If we consider the three-stage model of social norm formation, sentiment analysis is most prevalent during (1) attitude formation, and (2) behavioural reinforcement. Sentiment is tracked not through what people search per se, but rather what they are saying and commenting on.

A fundamental aspect of sentiment analysis is sentiment polarity, which refers to the classification of text along a continuum ranging from highly negative to highly positive. Sentiment polarity is typically measured on a numerical scale, where negative values represent negative sentiment, positive values indicate positive sentiment, and neutral values signify the absence of strong emotional expression. For example, a sentiment polarity score of -1.0 represents extreme negativity, while a score of $+1.0$ reflects extreme positivity. This makes polarity a quantitative spectrum where -0.9 is much more severe than -0.3, for example. However, while sentiment polarity provides a broad overview of emotional directionality, it does not capture the depth, intensity, or subjectivity of opinions, necessitating further refinement through additional sentiment analysis techniques.

Another key component of sentiment analysis is sentiment subjectivity, which differentiates between objective and subjective expressions within text. Objective statements may sometimes convey information that is verifiable and devoid of personal opinions, whereas subjective statements reflect personal beliefs, emotions, or evaluations. Subjectivity scores range from 0 (completely objective) to 1 (highly subjective), allowing researchers to determine the extent to which social media discussions are dominated by factual reporting versus opinionated discourse. Although there exist examples where highly subjective polarity scores overly high negativity, this is rare based on the lexical training of the models. By integrating subjectivity analysis with sentiment polarity detection, researchers can distinguish between emotionally charged but fact-based discussions and purely speculative, emotionally driven narratives.

5.4 NLP: Topic Modelling

Topic modelling is a method used in statistics and natural language processing (NLP) to uncover the main themes, that appear in a large collection of written texts. These collections include articles, social media posts, digital content, or online conversations like in social media. Topic modelling is especially useful for analysing unstructured text data, which is common in digital environments where people express themselves in free form. One of the strengths of topic modelling is that it doesn't require human guidance. At its core, topic modelling works by grouping words that tend to show up in similar contexts.

For example, if the words "vaccine" "virus," and "sickness" often appear together in many posts, the model might classify them as part of a "Infection" topic. The most used method is Latent Dirichlet Allocation (LDA), which assumes that each document (or message) discusses a mix of topics, and that each topic is made up of certain words with varying likelihoods. In LDA, the process involves assigning probabilities to words and topics. The algorithm cycles through the text, gradually improving its understanding of which words belong to which topics based on how often they appear together. This is done using a technique called Gibbs sampling, an iterative process that fine-tunes the model's predictions by constantly learning from the patterns in the data. The significance of topic modelling extends beyond technical applications. Topic modelling plays a key role in understanding how people make sense of the world, especially in the early stages of forming shared beliefs or social norms. In the context of social norm formation, sentiment alone and polarity alone cannot contextualise what is being discussed, and topic modelling adds an important layer to the analysis.

5.5 Social Stress Indicator

Introduced by Combrink [126], the SSI is a computational metric designed to quantify social stress levels in digital environments, particularly on social media platforms. It is built upon three key components, Sentiment Analysis (SA), Subjectivity (SUB), and Information-Seeking Behaviour (ISB). These components are integrated into a formula that enables real-time assessment of social stress levels by analysing online discourse. To use the indicator, a model must be trained. According to Combrink [126], the SSI is temporal and can be outlined as (Eq. (5.1)):

$$SSI = \frac{-SA + (1 - SUB) + ISB}{3}, \tag{5.1}$$

where *SA* indicates the sentiment analysis, *SUB* the subjectivity score, and *ISB* the information seeking behaviour of the posts. This indicator assumes an equal weighting, but given the nature of online discourse, Combrink [126] noted that a weighted SSI is needed to account for fluctuations in high SSI topics. When considering a scale free network, the SSI can serve as an indicator in the entire network that is tracked in a temporal manner. Integrating the SSI into a network model of online discourse presents a crucial advancement in Computational Infodemiology, because a more granular network can be modelled. There are a lot of research that still needs to be done to model the exact recommendations to these network responses, but this largely makes SSI in the context of Computational Infodemiology exciting.

5.6 Conclusion

This book has traced a new cartography of digital life, one where belief systems, once grounded in community, tradition, and stable epistemologies, are increasingly shaped by volatile information flows and algorithmic architectures. When sentiment fractures, when information-seeking spikes, when topics converge into unstable epistemic frames, we are not simply witnessing noise. We are witnessing stress, structured and quantifiable, and embedded within the circulatory system of modern belief.

The belief system maps developed in this work have revealed a profound insight, uncertainty is not merely a background condition, it is a structural force. Uncertainty undermines any attempt at fact-checking and sensationalises the potential spread of misinformation. In experimental simulations, we observed how the injection of uncertainty disrupts stable opinion clusters, catalyses polarisation, and reconfigures the pathways through which ideas gain traction. This realisation marks a turning point in Computational Infodemiology, uncertainty must be modelled not only as a variable, but as a topological agent within scale-free belief systems. Scale-free networks, modelled what is observed in digital echo-chambers. In such networks belief changes are cascading. The coupling of social stress with belief system structure is not optional, it is imperative, but a lot of work is needed to merge SSI with agents within a scale-free network. Without this integration, we risk misdiagnosing the triggers of collective behaviour, overlooking early signals of societal rupture, and failing to understand how belief resilience can be cultivated. This work offers not just a set of tools, but a call to reimagine how we measure, model, and ultimately mediate the stresses of knowing. We must move beyond static sentiment metrics toward dynamic, context-aware, networked interpretations of belief under pressure. As digital media continues to develop and envelop life, the field of Computational Infodemiology will become increasingly more important.

References

1. Abar, S., Theodoropoulos, G. K., Lemarinier, P., & O'Hare, G. M. (2017). Agent Based Modelling and Simulation tools: A review of the state-of-art software. *Computer Science Review, 24*, 13–33.
2. Albert, R., & Barabási, A. L. (2002). Statistical mechanics of complex networks. *Reviews of Modern Physics, 74*(1), 47.
3. Bouneffouf, D., Rish, I., & Aggarwal, C. (2020, July). Survey on applications of multi-armed and contextual bandits. In *2020 IEEE congress on evolutionary computation (CEC)* (pp. 1–8). IEEE.
4. Auer, P., Cesa-Bianchi, N., & Fischer, P. (2002). Finite-time analysis of the multiarmed bandit problem. *Machine Learning, 47*(2), 235–256.
5. Lin, L. J. (1992). Self-improving reactive agents based on reinforcement learning, planning and teaching. *Machine Learning, 8*(3), 293–321.
6. Pósfai, M., & Barabási, A. L. (2016). *Network science* (Vol. 3). Cambridge University Press.
7. Barabási, A. L., & Albert, R. (1999). Emergence of scaling in random networks. *Science, 286*(5439), 509–512.
8. Barto, A. G., & Mahadevan, S. (2003). Recent advances in hierarchical reinforcement learning. *Discrete Event Dynamic Systems, 13*(4), 341–379.
9. Baumann, F., Lorenz-Spreen, P., Sokolov, I. M., & Starnini, M. (2020). Modeling echo chambers and polarization dynamics in social networks. *Physical Review Letters, 124*(4), Article 048301.
10. Bicchieri, C. (2005). *The grammar of society: The nature and dynamics of social norms.* Cambridge University Press.
11. Bicchieri, C. (2016). *Norms in the wild: How to diagnose, measure, and change social norms.* Oxford University Press.
12. Biele, G., Rieskamp, J., & Gonzalez, R. (2009). Computational models for the combination of advice and individual learning. *Cognitive science, 33*(2), 206–242.
13. Bikhchandani, S., Hirshleifer, D., & Welch, I. (1992). A theory of fads, fashion, custom, and cultural change as informational cascades. *Journal of Political Economy, 100*(5), 992–1026.
14. Booysen, F., Guvuriro, S., & van Eyssen Combrink, H. M. (2024). Social discounting in a symmetric giving and taking frame: An artifactual field experiment with young South African adults. *Journal of the Experimental Analysis of Behavior, 122*(3), 322–334.
15. Bucher, T. (2019). The algorithmic imaginary: Exploring the ordinary effects of Facebook algorithms. In *The social power of algorithms* (pp. 30–44). Routledge.
16. Cairney, P., & Kwiatkowski, R. (2017). How to communicate effectively with policymakers: Combine insights from psychology and policy studies. *Palgrave Communications, 3*(1), 1–8.

H. Combrink, *Computational Infodemiology*,
SpringerBriefs in Computer Science, https://doi.org/10.1007/978-3-032-10130-3

17. Centola, D. (2018). *How behavior spreads: The science of complex contagions* (Vol. 3). Princeton University Press.
18. Zhang, X., Wang, J., Yu, C., Fei, J., Luo, T., & Cao, Z. (2025). Agent-based modeling of epidemics: Approaches, applications, and future directions. *Technologies, 13*(7), 272.
19. Checkland, P. (1981). *Systems thinking, systems practice.*
20. Cialdini, R. B., & Trost, M. R. (1998). Social influence: Social norms, conformity and compliance.
21. Cinelli, M., De Francisci Morales, G., Galeazzi, A., Quattrociocchi, W., & Starnini, M. (2021). The echo chamber effect on social media. *Proceedings of the National Academy of sciences, 118*(9), Article e2023301118.
22. Cinelli, M., Quattrociocchi, W., Galeazzi, A., Valensise, C. M., Brugnoli, E., Schmidt, A. L., Zola, P., Zollo, F., & Scala, A. (2020). The COVID-19 social media infodemic. *Scientific Reports, 10*(1), 16598.
23. Clauset, A., Shalizi, C. R., & Newman, M. E. (2009). Power-law distributions in empirical data. *SIAM Review, 51*(4), 661–703.
24. Colas, C., Sigaud, O., & Oudeyer, P. Y. (2018, July). Gep-pg: Decoupling exploration and exploitation in deep reinforcement learning algorithms. In *International conference on machine learning* (pp. 1039–1048). PMLR.
25. Franceschi, J., Pareschi, L., & Zanella, M. (2022). From agent-based models to the macroscopic description of fake-news spread: The role of competence in data-driven applications. *Partial Differential Equations and Applications, 3*(6), 68.
26. Zuardi, A. W., Crippa, J. A. S., Hallak, J. E. C., Pinto, J. P., Chagas, M. H. N., Rodrigues, G. G. R., Dursun, S. M., & Tumas, V. (2009). Cannabidiol for the treatment of psychosis in Parkinson's disease. *Journal of Psychopharmacology, 23*(8), 979–983.
27. Cornwall, A., & Jewkes, R. (1995). What is participatory research? *Social Science & Medicine, 41*(12), 1667–1676.
28. Daw, N. D., Gershman, S. J., Seymour, B., Dayan, P., & Dolan, R. J. (2011). Model-based influences on humans' choices and striatal prediction errors. *Neuron, 69*(6), 1204–1215.
29. Del Vicario, M., Bessi, A., Zollo, F., Petroni, F., Scala, A., Caldarelli, G., Stanley, H. E., & Quattrociocchi, W. (2016). The spreading of misinformation online. *Proceedings of the National Academy of Sciences, 113*(3), 554–559.
30. Donovan, J. (2020). Social-media companies must flatten the curve of misinformation. *Nature.*
31. Epstein, J. M. (1999). Agent-based computational models and generative social science. *Complexity, 4*(5), 41–60.
32. Erdős, P., & Rényi, A. (1960). On the evolution of random graphs. *Publications of the.*
33. Fazio, L. K. (2020). Repetition increases perceived truth even for known falsehoods. *Collabra: Psychology, 6*(1).
34. Finnemore, M., & Sikkink, K. (1998). International norm dynamics and political change. *International Organization, 52*(4), 887–917.
35. Floridi, L. (2016). Faultless responsibility: On the nature and allocation of moral responsibility for distributed moral actions. *Philosophical Transactions of the Royal Society A: Mathematical, Physical and Engineering Sciences, 374*(2083), 20160112.
36. Frank, M. J., Scheres, A., & Sherman, S. J. (2007). Understanding decision-making deficits in neurological conditions: Insights from models of natural action selection. *Philosophical Transactions of the Royal Society B: Biological Sciences, 362*(1485), 1641–1654.
37. Freire, P. (1970). *Pedagogy of the oppressed.* Continuum.
38. Fricker, M. (2007). *Epistemic injustice: Power and the ethics of knowing.* Oxford University Press.
39. Friston, K. J., Stephan, K. E., Montague, R., & Dolan, R. J. (2014). Computational psychiatry: The brain as a phantastic organ. *The Lancet Psychiatry, 1*(2), 148–158.
40. Wise, T., Zbozinek, T. D., Michelini, G., Hagan, C. C., & Mobbs, D. (2020). Changes in risk perception and self-reported protective behaviour during the first week of the COVID-19 pandemic in the United States. *Royal Society Open Science, 7*(9), Article 200742.

41. Gershman, S. J. (2015). A unifying probabilistic view of associative learning. *PLoS Computational Biology, 11*(11), Article e1004567.
42. Huys, Q. J., Maia, T. V., & Frank, M. J. (2016). Computational psychiatry as a bridge from neuroscience to clinical applications. *Nature Neuroscience, 19*(3), 404–413.
43. Gillespie, T. (2018). *Custodians of the internet: Platforms, content moderation, and the hidden decisions that shape social media.* Yale University Press.
44. Boyd, D., & Golebiewski, M. (2018). *Data voids: Where missing data can easily be exploited.*
45. Kridera, S., & Kanavos, A. (2024). Exploring trust dynamics in online social networks: A social network analysis perspective. *Mathematical and Computational Applications, 29*(3), 37.
46. Greenhalgh, T., Jackson, C., Shaw, S., & Janamian, T. (2016). Achieving research impact through co-creation in community-based health services: Literature review and case study. *The Milbank Quarterly, 94*(2), 392–429.
47. Gross, T., & Blasius, B. (2008). Adaptive coevolutionary networks: A review. *Journal of the Royal Society Interface, 5*(20), 259–271.
48. Ognyanova, K., Lazer, D., Robertson, R. E., & Wilson, C. (2020). Misinformation in action: Fake news exposure is linked to lower trust in media, higher trust in government when your side is in power. *Harvard Kennedy School Misinformation Review.*
49. Helberger, N., Karppinen, K., & D'acunto, L. (2018). Exposure diversity as a design principle for recommender systems. *Information, Communication & Society, 21*(2), 191–207.
50. Huys, Q. J., Pizzagalli, D. A., Bogdan, R., & Dayan, P. (2013). Mapping anhedonia onto reinforcement learning: A behavioural meta-analysis. *Biology of Mood & Anxiety Disorders, 3*(1), 12.
51. Islam, M. S., Sarkar, T., Khan, S. H., Kamal, A. H. M., Hasan, S. M., Kabir, A., Yeasmin, D., Islam, M. A., Chowdhury, K. I. A., Anwar, K. S., & Chughtai, A. A. (2020). COVID-19–related infodemic and its impact on public health: A global social media analysis. *The American Journal of Tropical Medicine and Hygiene, 103*(4), 1621.
52. Jackson, M. O. (2008). *Social and economic networks* (Vol. 3, p. 519). Princeton University Press.
53. Jackson, S. J., Bailey, M., & Welles, B. F. (2020). *#HashtagActivism: Networks of race and gender justice.* Mit Press.
54. Jaderberg, M., Mnih, V., Czarnecki, W. M., Schaul, T., Leibo, J. Z., Silver, D., & Kavukcuoglu, K. (2016). Reinforcement learning with unsupervised auxiliary tasks. arXiv:1611.05397
55. Jasanoff, S., & Simmet, H. R. (2017). No funeral bells: Public reason in a 'post-truth' age. *Social Studies of Science, 47*(5), 751–770.
56. Jasanoff, S. (2011). Cosmopolitan knowledge: Climate science and global civic epistemology.
57. Kable, J. W., & Glimcher, P. W. (2007). The neural correlates of subjective value during intertemporal choice. *Nature Neuroscience, 10*(12), 1625–1633.
58. Kaelbling, L. P., Littman, M. L., & Moore, A. W. (1996). Reinforcement learning: A survey. *Journal of Artificial Intelligence Research, 4*, 237–285.
59. Keeling, M. J., & Eames, K. T. (2005). Networks and epidemic models. *Journal of the Royal Society Interface, 2*(4), 295–307.
60. Kermack, W. O., & McKendrick, A. G. (1927). A contribution to the mathematical theory of epidemics. *Proceedings of the Royal Society of London. Series A, Containing Papers of a Mathematical and Physical Character, 115*(772), 700–721.
61. Di Domenico, G., Sit, J., Ishizaka, A., & Nunan, D. (2021). Fake news, social media and marketing: A systematic review. *Journal of Business Research, 124*, 329–341.
62. Kwak, H., Lee, C., Park, H., & Moon, S. (2010, April). What is Twitter, a social network or a news media?. In *Proceedings of the 19th International Conference on World Wide Web* (pp. 591–600).
63. Ammara, U., Bukhari, H., & Qadir, J. (2020, October). Analyzing misinformation through the lens of systems thinking. In *TTO* (pp. 55–63).
64. Wiegand, M., Eagan, R. L., Karimov, R., Lin, L., Larson, H. J., & Figueiredo, A. D. (2023). Global declines in vaccine confidence from 2015 to 2022: A large-scale retrospective analysis. SSRN 4438003.

65. Fairhead, J., & Leach, M. (2012). *Vaccine anxieties: Global science, child health and society.* Routledge.
66. Legros, S., & Cislaghi, B. (2020). Mapping the social-norms literature: An overview of reviews. *Perspectives on Psychological Science, 15*(1), 62–80.
67. Lerman, K., & Ghosh, R. (2010, May). Information contagion: An empirical study of the spread of news on digg and twitter social networks. In *Proceedings of the International AAAI Conference on web and Social Media* (Vol. 4, No. 1, pp. 90–97).
68. Levine, S., Pastor, P., Krizhevsky, A., Ibarz, J., & Quillen, D. (2018). Learning hand-eye coordination for robotic grasping with deep learning and large-scale data collection. *The International Journal of Robotics Research, 37*(4–5), 421–436.
69. Lewandowsky, S., Ecker, U. K., & Cook, J. (2017). Beyond misinformation: Understanding and coping with the "post-truth" era. *Journal of Applied Research in Memory and Cognition, 6*(4), 353–369.
71. Macal, C. M., & North, M. J. (2005, December). Tutorial on agent-based modeling and simulation. In *Proceedings of the Winter Simulation Conference, 2005.* (pp. 14–pp). IEEE.
72. Madumal, P., Miller, T., Sonenberg, L., & Vetere, F. (2020, April). Explainable reinforcement learning through a causal lens. In *Proceedings of the AAAI Conference on Artificial Intelligence* (Vol. 34, No. 03, pp. 2493–2500).
73. Bottle, A., & Aylin, P. (2011). Predicting the false alarm rate in multi-institution mortality monitoring. *Journal of the Operational Research Society, 62*(9), 1711–1718.
74. Midgley, G. (2003). Systemic intervention: Philosophy, methodology, and practice. *Journal of Community and Applied Psychology, 13*(4), 330–333.
75. Midgley, G. (2006). Systems thinking for evaluation. In *Systems concepts in evaluation: An expert anthology* (pp. 11–34).
76. Midgley, G., Cavana, R. Y., Brocklesby, J., Foote, J. L., Wood, D. R., & Ahuriri-Driscoll, A. (2013). Towards a new framework for evaluating systemic problem structuring methods. *European Journal of Operational Research, 229*(1), 143–154.
77. Mislove, A., Marcon, M., Gummadi, K. P., Druschel, P., & Bhattacharjee, B. (2007, October). Measurement and analysis of online social networks. In *Proceedings of the 7th ACM SIGCOMM Conference on Internet Measurement* (pp. 29–42).
78. Jarrahi, M. H., Newlands, G., Lee, M. K., Wolf, C. T., Kinder, E., & Sutherland, W. (2021). Algorithmic management in a work context. *Big Data & Society, 8*(2), 20539517211020332.
79. Mkungeka, P., Combrink, H. M., & de Wet, K. (2025). South African social media perspectives on COVID-19 during 2021. *Journal of Consumer Health on the Internet,* 1–16.
80. Mnih, V., Kavukcuoglu, K., Silver, D., Rusu, A. A., Veness, J., Bellemare, M. G., Graves, A., Riedmiller, M., Fidjeland, A. K., Ostrovski, G., & Petersen, S. (2015). Human-level control through deep reinforcement learning. *Nature, 518*(7540), 529–533.
81. Huys, Q. J. (2013). Computational psychiatry. In *Encyclopedia of computational neuroscience* (pp. 1–10). Springer.
82. Moody, J., & Saffell, M. (2001). Learning to trade via direct reinforcement. *IEEE Transactions on Neural Networks, 12*(4), 875–889.
83. Chenzi, V. (2021). Fake news, social media and xenophobia in South Africa. *African Identities, 19*(4), 502–521.
84. Diedong, A. L. (2020). Vibrant and safe media landscape in Ghana: Reality or mirage? *Journal of African Media Studies, 12*(2), 157–170.
85. North, M. J., & Macal, C. M. (2007). *Managing business complexity: Discovering strategic solutions with agent-based modeling and simulation.* Oxford University Press.
86. Nussbaum, M. C. (2007). Capabilities as fundamental entitlements: Sen and social justice. In *Capabilities equality* (pp. 54–80). Routledge.
87. Nussbaum, M. C. (2007). *Frontiers of justice: Disability, nationality, species membership.* Harvard University Press.
88. Nussbaum, M. C. (2011). *Creating capabilities: The human development approach.* Harvard University Press.

89. Boaz, A., & Davies, H. (Eds.). (2019). *What works now?: Evidence-informed policy and practice*. Policy press.
90. Abeille, M. (2017). *Exploration-exploitation with Thompson sampling in linear systems* (Doctoral dissertation, Université de Lille 1).
91. Pastor-Satorras, R., & Vespignani, A. (2001). Epidemic spreading in scale-free networks. *Physical Review Letters, 86*(14), 3200.
92. Pavlov, P. I. (2010). Conditioned reflexes: An investigation of the physiological activity of the cerebral cortex. *Annals of Neurosciences, 17*(3), 136.
93. Pennycook, G., & Rand, D. G. (2021). The psychology of fake news. *Trends in Cognitive Sciences, 25*(5), 388–402.
94. Prentice, D. A., & Miller, D. T. (1993). Pluralistic ignorance and alcohol use on campus: Some consequences of misperceiving the social norm. *Journal of Personality and Social Psychology, 64*(2), 243.
95. Puterman, M. L. (2014). *Markov decision processes: Discrete stochastic dynamic programming*. Wiley.
96. Rand, W., & Rust, R. T. (2011). Agent-based modeling in marketing: Guidelines for rigor. *International Journal of research in Marketing, 28*(3), 181–193.
97. Ravasz, E., & Barabási, A. L. (2003). Hierarchical organization in complex networks. *Physical Review E, 67*(2), Article 026112.
98. Reynolds, M., & Holwell, S. (Eds.). (2020). *Systems approaches to making change: A practical guide* (Vol. 2). Springer London.
99. Schultz, W., Dayan, P., & Montague, P. R. (1997). A neural substrate of prediction and reward. *Science, 275*(5306), 1593–1599.
100. Sen, A. (2009). *The idea of justice*. Harvard University Press.
101. Shao, C., Ciampaglia, G. L., Varol, O., Flammini, A., & Menczer, F. (2017). The spread of fake news by social bots, *96*(104), 14. arXiv:1707.07592
102. Sîrbu, A., Loreto, V., Servedio, V. D., & Tria, F. (2016). Opinion dynamics: Models, extensions and external effects. *Participatory sensing, opinions and collective awareness* (pp. 363–401). Springer International Publishing.
103. Gilbert, S. A. (2020). "I run the world's largest historical outreach project and it's on a cesspool of a website." Moderating a Public Scholarship Site on Reddit: A Case Study of r/AskHistorians. *Proceedings of the ACM on Human-Computer Interaction, 4*(CSCW1), 1–27.
104. Sunstein, C. R. (2020). *Too much information: Understanding what you don't want to know*. MIT Press.
105. Sutton, R. S., & Barto, A. G. (1998). *Reinforcement learning: An introduction* (Vol. 1, No. 1, pp. 9–11). MIT press.
106. Wang, X. J., & Krystal, J. H. (2014). Computational psychiatry. *Neuron, 84*(3), 638–654.
107. Tankard, M. E., & Paluck, E. L. (2016). Norm perception as a vehicle for social change. *Social Issues and Policy Review, 10*(1), 181–211.
108. Taylor, M. E., & Stone, P. (2009). Transfer learning for reinforcement learning domains: A survey. *Journal of Machine Learning Research, 10*(7).
109. Arnaout, R., & Arnaout, R., 2022. Visualizing omicron: COVID-19 deaths versus cases over time. *Plos One, 17*(4), e0265233.
110. Tucker, J. A., Guess, A., Barberá, P., Vaccari, C., Siegel, A., Sanovich, S., Stukal, D., & Nyhan, B. (2018). Social media, political polarization, and political disinformation: A review of the scientific literature. In *Political polarization, and political disinformation: A review of the scientific literature*.
111. Ulrich, W. (2003). Beyond methodology choice: Critical systems thinking as critically systemic discourse. *Journal of the Operational Research Society, 54*(4), 325–342.
112. Van Dijck, J., Poell, T., & De Waal, M. (2018). *The platform society: Public values in a connective world*. Oxford University Press.
113. Vosoughi, S., Roy, D., & Aral, S. (2018). The spread of true and false news online. *Science, 359*(6380), 1146–1151.
114. Watkins, C. J., & Dayan, P. (1992). Q-learning. *Machine Learning, 8*(3), 279–292.

115. Watts, D. J., & Strogatz, S. H. (1998). Collective dynamics of 'small-world' networks. *Nature, 393*(6684), 440–442.
116. Wooldridge, M. (2009). *An introduction to multiagent systems.* Wiley.
117. World Health Organization. (2020). Managing the COVID-19 infodemic: Promoting healthy behaviours and mitigating the harm from misinformation and disinformation.
118. Yom-Tov, E., Feraru, G., Kozdoba, M., Mannor, S., Tennenholtz, M., & Hochberg, I. (2017). Encouraging physical activity in patients with diabetes: Intervention using a reinforcement learning system. *Journal of Medical Internet research, 19*(10), Article e338.
119. Zollo, F., Novak, P. K., Del Vicario, M., Bessi, A., Mozetič, I., Scala, A., Caldarelli, G., & Quattrociocchi, W. (2015). Emotional dynamics in the age of misinformation. *Plos One, 10*(9), Article e0138740.
120. Törnberg, P. (2022). How digital media drive affective polarization through partisan sorting. *Proceedings of the National Academy of Sciences, 119*(42), e2207159119.
121. Midgley, G., & Pinzón, L. A. (2011). Boundary critique and its implications for conflict prevention. *Journal of the Operational Research Society, 62*(8), 1543–1554.
122. Fairhead, J., & Leach, M. (2007). Partnerships for sustainable timber production in Trinidad: dealing with social and ecological dynamics. Partnerships in sustainable forest resource management: learning from Latin America. Amsterdam, NL. *CEDLA Latin America Studies.* 107–124.
123. Mackie, G., Moneti, F., Shakya, H., & Denny, E. (2015). What are social norms? How are they measured. *University of California at San Diego-UNICEF Working Paper, San Diego.*
124. Gershman, S. J., & Daw, N. D. (2017). Reinforcement learning and episodic memory in humans and animals: an integrative framework. *Annual review of psychology, 68*(1), 101–128.
125. Pearl, J. (2009). Causal inference in statistics: An overview.
126. Combrink, H. (2025). The Social Stress Indicator (SSI). *PLoS One, 20*(9), e0328768.

MIX
Papier aus verantwortungsvollen Quellen
Paper from responsible sources
FSC® C105338

If you have any concerns about our products,
you can contact us on
ProductSafety@springernature.com

In case Publisher is established outside the EU,
the EU authorized representative is:
Springer Nature Customer Service Center GmbH
Europaplatz 3, 69115 Heidelberg, Germany

Printed by Libri Plureos GmbH
in Hamburg, Germany